Manufacturing Technology and Design

Manufacturing Technology and Design

Anurag Shrivastava

Manufacturing Technology and Design

ISBN 978-93-5111-411-6

Published in 2014 in India by

RANDOM PUBLICATIONS

4376-A/4B, Gali Murari Lal, Ansari Road
New Delhi-110 002
Phone : +91-11-43580356, +91-11-23289044
e-mail: randomexports@gmail.com, sales@randompublications.com, info@randompublications.com

Reprinted 2021

Type Setting by : Keystoneprintads, Delhi-110051
Digitally Printed at: Replika Press Pvt. Ltd.

Preface

The Manufacturing Technology is the areas of machining, drafting and fluid systems in order to prepare them for entry-level positions as machine operators, machine maintenance personnel and quality control personnel. Manufacturing technology provides the tools that enable production of all manufactured goods. Manufacturing technology provides the tools that enable production of all manufactured goods. These master tools of industry magnify the effort of individual workers and give an industrial nation the power to turn raw materials into the affordable, quality goods essential to today's society. In short, we make modern life possible.

Manufacturing technology provides the productivity tools that power a growing, stable economy and a rising standard of living. These tools create the means to provide an effective national defense. They make possible modern communications, affordable agricultural products, efficient transportation, innovative medical procedures, space exploration... and the everyday conveniences we take for granted.

Advanced manufacturing technologies (AMTs) involve new manufacturing techniques and machines combined with the application of information technology, microelectronics and new organizational practices within the manufacturing process. These hardware technologies have found wide acceptance in discrete manufacturing and in resource and processing sectors. Computer numeric controls have been applied to systems for machining, forming, cutting and molding.

Today's product designers are presented with a myriad of choices when creating their work and preparing it for manufacture. They have to be knowledgeable about a vast repertoire of processes, ranging from what used to be known as traditional "crafts" to the latest technology, to enable their designs to be manufactured effectively and efficiently. Information on the internet about such processes is often unreliable, and search engines do not usefully organize material for designers.

The book covers manufacturing processes of extruded and moulded products with the various mould designs.

I thank all members of my team who have helped in the preparation of the book. My special thanks go to "Random Publications" who have published the book.

– ***Anurag Shrivastava***

Contents

1

Introduction of Advanced Manufacturing Technology

INTRODUCTION

The market place of the twenty-first century is evolving into one of merging national markets, fragmented consumer markets, and rapidly changing product technologies. These changes are driving firms to compete, simultaneously, along several different dimensions: design, manufacturing, distribution, communication, sales and others. Although manufacturing has not been utilized as a competitive weapon historically, the market place of the twenty first century will demand that manufacturing assume a crucial role in the new competitive arena.

Progress in human society has been accomplished by the creation of new technologies. The last few years have witnessed unparalleled changes throughout the world. Rapid changes in the markets demand drastically shortened product life cycles and high-quality products at competitive prices. Customers now prefer a large variety of products. This phenomenon has inspired manufacturing firms to look for progressive computerized automation in various processes. Thus mass production is being replaced by low-volume, high-variety production. Manufacturing firms have recognized the importance of flexibility in the manufacturing system to meet the challenges posed by the pluralistic market. The concept of flexibility in manufacturing systems has attained significant importance in meeting the challenges for a variety of products of shorter lead-times, together with higher productivity and quality. The flexibility is the underlying concept behind the transition from traditional methods of production to the more automated and integrated methods. They stress that firms implementing automation projects should prioritize their needs for different flexibilities for long-range strategic perspectives.

Intensifying global competition and rapid advancement of manufacturing technology are two realties in today's business environment. These have combined to shift the business strategic priorities toward quality, cost effectiveness and responsiveness to marketplace changes. The quest for lower operating costs and improved manufacturing efficiency has forced a large number of manufacturing firms to embark on AMTs projects of various types. Dramatic developments in AMT at various organizational levels can be attributed to numerous benefits that improve the competitive position of the adopting companies. AMT impact not just manufacturing, but the whole business operations, giving new challenges to a firm's ability to mange both manufacturing and information technologies.

AMT can also provide distinctive competitive advantages in cost and process leadership. Events of the last decade, such as the US productivity problems, Japanese manufacturing success stories and the competitive global economy, have moved manufacturing strategy and process technology issues from the bottom to the top of the firm's priority list. The issues surrounding manufacturing technologies and their implementations have assumed greater importance in the manufacturing strategy debate. Practitioners and researchers have developed strong interest in how AMT can be used as a competitive tool in the global economy. A growing number of organizations are now adopting AMT to cope with recent phenomena in today's competitive environment such as fragmented mass markets, shorter product life cycle and increased demand for customization. Although AMT can help manufacturers compete under these challenging circumstances, they often serve as a double-edged sword, imposing organizational challenges while providing distinct competitive advantage when successfully implemented.

International business strategies frequently demand the transfer of manufacturing processes. Manufacturing process is defined as any repetitive system for producing a product, including the people, equipment, material inputs, procedures and software in that system. An issue of importance in this strategic decision is whether the process should be transferred without modification or adopted in some way for transfer.

Owing to the intense global competition in manufacturing, manufacturers need to increase their level of competitiveness in the global market. Some manufacturing companies, therefore, are forced to undergo a period of transformation in order to compete more effectively. Under these circumstances, AMT is considered as a means of improving competitiveness.

The term AMT refers to computer-aided technologies in design, manufacturing, transportation and testing, etc. In general, AMT can be categorized into two principle ways:

- The classical continuum of basic manufacturing processes which extends from make-to-order manufacturing to continuous manufacturing; and the level of integration of the overall manufacturing system.
- AMT provides an organization with an opportunity to successfully combat market place dynamics and create for itself a competitive

advantage. Manufacturers and academics believe that AMTs can reduce operating cots, provide high levels of output by removing inconsistent human input, improve manufacturing flexibility and lead time to market.

The literature on Advanced Manufacturing Technologies can be divided in to different fields but all fields are interrelated and those fields are as follows:

- Adoption of Advanced manufacturing technologies
- Investment of AMT
- Selecting and Assessment of AMT
- Planning and Implementation of AMTs
- Historical development of AMT
- Definitions and Group of AMTs
- Benefits and Disadvantages of AMT

DEFINITIONS OF AMT

Numerous definitions of AMT exist. For example, Baldwin (1995) defines AMT as 'a group of integrated hardware-based and software-based technologies, which if properly implemented, monitored, and evaluated, will lead to improving the efficiency and effectiveness of the firm in manufacturing a product or providing a service.' This work, however, provides a more global definition that combines the work of Zhao, and Zammuto and O'Connor (1995). AMT, defined broadly, is a total socio-technical system where the adopted methodology defines the incorporated level of technology. AMT employs a family of manufacturing (CAM), flexible manufacturing systems (FMS), manufacturing resource planning (MRP II), automated material handling systems, robotics, computer-numerically controlled (CNC) machines, computer-integrated manufacturing (CIM) systems, optimized production technology (OPT), and just-in-time (JIT). Although AMT places great emphasis on the use of technological innovation, management's role is significant since AMT systems require continual review and readjustment.

Advanced Manufacturing technology (AMT) represents a wide variety of mainly computer-based systems, which provide adopting firms with the potential to improve manufacturing operations greatly. It is generally expected that the resultant improvement in operational performance will enhance the firm's ability to reap the underlying marketing, strategic and business benefits for which the systems were adopted. Another definition is AMT refers to a family of technologies that include computer-aided design (CAD) and engineering systems, materials resource planning systems, automated materials handling systems, robotics, computer controlled machines, flexible manufacturing systems, electronic data interchange and computer-integrated manufacturing systems.

AMT represents a wide variety of modern computer-based systems devoted to the improvement of manufacturing operations and thereby enhancement of firm competitiveness. Global competition continues to drive the adoption of advanced manufacturing technology (AMT) to date, the literature AMT on the introduction of AMT new technology is based primarily on case studies and expert opinions.

AMT has been described as programmable machinery or a system of programmed machines that can produce a variety of products or parts with virtually no time lost for changes. The process of introducing AMT is generally evolutionary but in some settings the pace is quickening. As innovations spill over into practical application there is a simultaneous change in the mix and range of elements generally labeled as AMT. The machines, computer control and linkages, as well as the human operator involvement, appear to be on a windy route from islands of technology towards some far wider degree of computer integration, referred to as computer integrated manufacturing (CIM).

AMT represents a wide variety of modern computer-based or numerical control-based systems devoted to the improvement of manufacturing operations. AMT is broadly defined as 'an automated production system of people, machines and tools for the planning and control of the production process, including the procurement of raw materials, parts and components and the shipment and service of finished products. The properties of AMT overcome the limitations of conventional technology in enabling small firms to develop economies of scope based on low volume and low cost production. Specifically, AMT facilitates customization and reduces lead times through the productions of variety, frequent design changeovers, and rapid processing of design, assembly, materials handling and market information.

In general, AMT typically involves (a) a computer-aided design system (CAD) that develops designs, displays them and stores them for future reference: (b) a computer-aided manufacturing system (CAM) that translates CAD information for production and further controls machine tools, material flow, and testing; (c) an automotive storage and retrieval system for delivery or pack up of parts between machines and storage: and (d) a supervisory computer that integrates all of the above (CIM).

AMTs refer to a family of technologies that include computer-aided design (CAD) and engineering systems, materials resource planning systems, automated materials handling systems, robotics, computer-controlled machines, flexible manufacturing systems, electronic data interchange and computer-integrated manufacturing systems.

GROUP OF AMT

Advanced manufacturing Technologies involve the application of computers to various facets of the production process. The 22 manufacturing technologies are grouped into six functional categories, each capturing a different aspect of the process-fabrication and assembly, automated materials handling systems, design and engineering, inspection and communications, manufacturing information systems, or integration and control. There are 26 AMTs listed in Baldwin's (1995) survey. The list is obtained from Statistics Canada in 1995. The 26 technologies belong to six functional technology groups-design and engineering; processing, fabrication, and assembly, automated material handling, inspection, network communications, and integration and control.

Advanced manufacturing technologies (AMTs) involve new manufacturing techniques and machines combined with the application of information technology, microelectronics and new organizational practices within the manufacturing process. These hardware technologies have found wide acceptance in discrete manufacturing and in resource and processing sectors. Computer numeric controls have been applied to systems for machining, forming, cutting and molding. The accuracy, speed and control of robots have improved significantly and, as a result, they are used extensively in welding, painting, material handling and an enormous number of unique assembly applications. Vision systems provide, in real time, monitoring for precision machining and high-speed printing and remote handling of mining equipment used thousands of Metrecs underground. Computer-aided design (CAD) and rapid prototyping have substantially shortened the development time for new products. Computer-integrated manufacturing (CIM) is applied in machine shops for tool building, the production of engines and body assemblies for passenger cars, the manufacture of airplane landing gear and the production of hypodermic needles for the medical devices sector.

Complementary to the hardware technologies of AMTs is a wide range of 'soft' manufacturing process technologies. Just-in-time (JIT) manufacturing, total quality management (TQM) and supply chain management are but a few of the many 'soft' AMTs adopted by manufacturers and processors globally. The nature of competition in manufacturing has changed. High-quality and highly customized goods are demanded. There is a premium for being the first to market with a product. This has created a demand for 'hard' and 'soft' technologies that help shorten design and production cycles. Freer trade has increased the breadth of geographical competition, making it easy for foreign manufacturers to enter the Sri Lankan market and for Sri Lankan firms to enter foreign markets. The source of this competition varies, some of it from low-wage areas and some of it from very technologically sophisticated countries.

Manufacturing practices and processes have come under increased pressure from global competition. Demands for improved customer service, breadth of product line, improved quality, quick response and a much shortened time-to-market for new product introduction cannot be ignored by firms. In the face of these intense pressures, Sri Lankan and other manufacturers around the world are moving away from mass production manufacturing processes. They are turning to greater flexibility and speed in manufacturing practices. These practices have become the foundation for 'Best in Class' manufacturers and processors. Manufacturing has been influenced by trade liberalization, global competition, market fragmentation, technological innovation and the demands of more sophisticated consumers. In response to these pressures, manufacturers are incorporating more flexibility and technology in their production practices. These features have become a trademark of world-class corporations. AMTs are a key enabler for firms attempting to meet world-class performance targets.

Table. Type of Advanced Manufacturing Technologies

Functional Group	Technology
Processing, Fabrication and Assembly	Flexible manufacturing cells or systems (FMC/FMS) Programmable logic control machines or processes (CNC and NC) Lasers used in materials processing Robots with sensing capabilities Robots without sensing capabilities Rapid Prototyping systems High speed machining Near net shape technologies
Automated Material Handling	Part identification for manufacturing automation (bar coding) Automated storage and Retrieval system (AS/RS) Automated Guided Vehicle Systems (AGVS)
Design and Engineering	Computer-aided design and engineering (CAD/CAE) Computer-Aided Design/Manufacturing CAD/CAM Modeling or simulation Technologies Electronic exchange of CAD files Digital representation of CAD output
Inspection and Communications	Automated Vision-based systems used for inspection/testing of inputs/or final products Other automated sensor-based systems used for inspection/testing of inputs
Manufacturing Information Systems	Manufacturing Resource Planning (MRP) Manufacturing Resource Planning (MRP11)
Integration and Control	Supervisory Control & Data Acquisition (SCADA) Artificial Intelligence/Expert Systems (AI) Computer Integrated Manufacturing (CIM)

The functional groups and their constituent technologies, along with a brief description of each, are provided in Table. AMT is a generic term for a group of manufacturing technologies, which combine both scope and scale capabilities in a manufacturing environment. Manufacturing strategy has become more sophisticated. As a result, AMT can play a crucial role in making it possible for firms to compete on 'traditionally' contradictory competitive priorities simultaneously.

Baldwin (1995) suggests two subgroups of technologies within AMT: the traditional hardware technology consisting of systems, devices and stations (SDS): and a second group of technologies, often in software form, which perform integrative and managerial functions-integrative and managerial systems (IMS).

Typical examples of systems, devices, and stations (SDS) include automated identification station, automated inspection stations, automated material handling devices, computer aided design workstations, computerized numerical control machine tools, numerical control machine tools, programmable production controllers, robots, shop-floor control systems.

Examples of integrated and managerial systems (IMS) include computer aided manufacturing, computer-aided engineering, statistical process control, production planning/inventory management software, engineering data management, computer aided process planning, local area networks, group technology.

Both SDS and IMS technologies can be used individually or in combination with other technologies to achieve desired economies of scale and scope. When taken together SDC and IMS constitute AMT.

AMT represents a wide variety of mainly computer-based systems that provide adopting firms with the potential to improve manufacturing operations greatly. It is generally expected that the resultant improvement in operational performance will enhance the firm's ability to reap the underlying marketing, strategic and business benefits for which the systems were adopted. Some of the benefits attributed to these technologies are improving market share, gaining earlier entrance to market share, responding more quickly to changing customer needs and the quality to offer products with improved quality and reliability.

These technologies have been classified as stand-alone systems, intermediate systems and integrated systems. Technologies such as computer aided design (CAD) and computer numerical control machines (CNC) are typically categorized as standalone systems. Automated material handling systems (AMHS) and automated inspection and testing systems (AITS) are classified as intermediated systems. Integrated technologies can be categorized as either integrated process technologies (e.g. computer manufacturing systems (CIM) and flexible manufacturing systems (FMS) or integrated information/logistic technologies (e.g. just-in-time production (JIT) and manufacturing resources planning (MRPII). The general trend in the AMT research literature has been to examine technology adoption and any resultant change in firm performance on the basis of the implementation of individual technologies or in terms of implementation of specific technology classifications such as those presented above.

AMTs refer to a family of technologies that include computer-aided design (CAD) and engineering systems, materials resource planning systems, automated materials handling systems, robotics, computer controlled machines, flexible manufacturing systems, electronic data interchange and computer-integrated manufacturing systems (Zammuto and O'Connor, 1992). Numerous studies have emphasized the potential strategic benefits of flexibility, responsiveness, improved quality and improved productivity through purposeful investment in AMTs. Such benefits are increasingly important in the current global manufacturing environment, which has been described as hyper competitive, high-velocity and characterized by fragmenting markets, shorter product life cycles, and increasing consumer demand for customization (Zammuto and O'Connor, 1992). Thus, AMTs have, and will continue to have, a key strategic role in improving competitiveness by utilizing the manufacturing function more effectively in overall business strategy.

AMT includes a group of integrated hardware-based and software-based technologies which, when properly implemented, monitored and evaluated, can improve the operating efficiency and effectiveness of the adopting firms. They encompass a broad range of computer-based technological innovations, which include numerical control (NC) machine tools, cellular manufacturing, machining (CAD/CAM) systems, and automated storage and retrieval systems (AS/RS). These islands of automation are integrated manufacturing (CIM). AMT has the potential to improve operating performance dramatically and create vital business opportunities for companies, which are capable of successfully implementing and managing them.

AMT can also provide distinctive competitive advantages in cost and process leadership. Events of the last decade, such as the US productivity problems, Japanese manufacturing success stories and the competitive global economy, have moved manufacturing strategy and process technology issues from the bottom to the top of the firm's priority list.

The AMTs are broadly classified into seven sub-groups. The classification scheme adopted here is similar to the US Department of Commerce Survey of Manufacturing Technology. The findings of some researchers show that the technologies are cross categorized as stand-alone systems and integrated systems.

This classification scheme links technologies that have similar benefits and costs. The 14 technologies are classified into three main groups, and seven sub-groups are further divided as shown in Table below.

Stand-alone systems

- Design and engineering technologies
  - Computer-aided design (CAD)
  - Computer-aided process planning (CAPP)
- Fabricating/machine and assembly technologies
  - NC/CNC or DNC machines
  - Materials working laser (MWL)
  - Pick-and –place robots
  - Other robots
  - Intermediate systems
- Automated material handling technologies
  - Automatic storage and retrieval systems (AS/RS)
  - Automated material handling systems (AMHS)
- Automated inspection and testing systems
  - Automated Inspection and testing equipment (AITE)
  - Integrated systems
- Flexible manufacturing technologies
  - Flexible manufacturing cells/systems (FMC/FMS)
- Computer-integrated manufacturing systems
  - Computer-integrated manufacturing (CIM)
- Logistic related systems
  - Just-in-time (JIT)
  - Material requirements planning
  - Manufacturing resources planning (MRPII)

Companies must ascertain which technologies can fulfill their objectives and identify the selected technologies belonging to a system, since it will affect the following justification methods chosen. For stand-alone systems where the purpose is the straightforward replacement of old equipment, even if some economic benefits not usually considered are obtained, the standard economic justification approaches can be used.

There is another grouping method based on the amount or type of investment. Nature of the AMT investment can be divided into three-section computer hardware, computer software and plant and equipment.

- AMT related hard ware:
  - LAN (local area networks)
  - Micros (PC Computers)
  - Graphics hardware:
  - Mainframe
  - Online process instrumentation
  - Shop floor data capture
  - WAN (wide area networks)
- AMT related Software:
  - CAD/CAM

  - Data base management systems
  - MRPII (Manufacturing Resource planning I and II)
- QC software
  - MRP
  - Expert systems
  - OPT
  - MAP
- AMT related plant and equipment:
  - CNC M/Cs
  - Automatic testing equipment
  - Computer-controlled testing equipment
  - Flexible manufacturing cells
  - Automatic assembly
  - Flexible assembly systems
  - Robots
  - Laser cutting
  - Automated warehousing/order picking
  - AGVs
  - Laser measuring

Generic Advanced Manufacturing Technologies and Practices is another AMT grouping. Generic AMT represent the different generations and different capabilities that crosscut most industries.

- Multi-axes machining center
- High speed machining center
- Cellular manufacturing or group layout
- Computerized production planning system
- Product modeling and simulation
- Electronic exchange of product design and process data
- Automated parts identification devices
- Rapid set tooling and fixturing
- Rapid prototyping
- Computerized process planning system
- Product data management system
- Process simulation
- Knowledge-based systems
- On-line intelligent process control

Advanced manufacturing technologies refer to a family of technologies used in all facets of manufacture including design, control, fabrication and assembly. AMTs have taken on many forms of acronyms, including computer-aided design (CAD), computer aided engineering (CAE), manufacturing resource planning (MRPII), automated materials handling system, electronic data interchange (EDI), computer integrated manufacturing (CIM) systems and flexible manufacturing cell (FMC).

Empirical Evidence of Productivity increases of the firm

Sufficient evidence exists that computer aided technologies can increase productivity as well as quality and production flexibility. In a study of 14 organizations that have deployed computer-aided manufacturing technologies, Voss found that 12 had experienced increases in productivity. Increased product quality, reduced lead-times, and greater flexibility occurred in 8 of them. On the other hand, a pioneer study by Jaikumar (1986) found that deployment of computer-aided technologies in some firms only boosted machine up time and productivity or lead-times. Jaikumar concluded that these organizations did not realize increase in quality and flexibility because they superimposed the new technologies on existing structured organizations. Gaining the inherent flexibility and quality advantages of computer-aided technologies requires a more integrated and flexible work organization with decentralized control structures.

Computer-aided technologies are similar to earlier conventional non-programmable fixed-cycle technologies because of their ability to increase machine up time and produce high volume products. Scale economies needed to be competitive in mass markets can thus be reaped with computer-aided technologies. Nonetheless, these new technologies differ from conventional machines because they have the capacity to produce a wide array of low volume, high quality, customized parts or products needed for a highly segregated market (Zammuto and O'Connor, 1992).

The issues surrounding manufacturing technologies and their implementations have assumed greater importance in the manufacturing strategy debate. Practitioners and researchers have developed strong interest in how AMT can be used as a competitive tool in global economy. A growing number of organizations are now adopting AMT to cope with recent phenomena in today's competitive environment such as fragmented mass markets, shorter product life cycle and increased demand for customization. AMT can help serve as a double-edged sword, imposing organizational challenges while providing distinct competitive advantage when successfully implemented.

History of AMT

Historically, manufacturing organizations made improvements in their productions processes primarily through investments in physical capital. Advices in mechanization, for example, enabled manufacturing firms to enhance efficiency in production while actually lowering the required skills and capabilities of employees. As a result, many firms actually took the tactic of deskilling their workforces to reduce labor expenses, thereby diminishing the necessary level of investments in human capital with the hope of increasing profits. In contrast to this perspective, advocates of the cotemporary manufacturing paradigm argue for a dramatically different orientation toward employees. These post-industrial theorists propose that compared to the deskilling tactics of traditional manufacturing, more advanced systems require a set of complementary practices for upskilling the workforce. Indeed, some studies have suggested that modern manufacturing systems demand more enhanced technical, conceptual, analytic and problem solving skills than typically were required in traditional manufacturing environments.

In the manufacturing sector, the new technological advances have revolved around machine tools and equipment. Advances in automation of machine tools began about 30 years ago with the first generation of programmable machine tools-the numerical controlled machine tools (NC). The NC did not become widely used until the 1970s. The second generation, computerized numerical control machine (CNC), was introduced in the late 1970s. These new machine tools have the capacity to produce the high volume standardized parts and products necessary for competitive success in undifferentiated markets.

The first AMT technologies were introduced in the 1950s, but it was not until the 1970s that the adoption of AMTs took off and the 1980s that their use became widespread. Today, nearly all currently produced manufacturing equipment incorporates some electronics element and thus fits the definitions for AMTs.

There are three types of manufacturing systems: crafts shops; dedicated manufacturing systems (DMS); and advanced manufacturing technology – based systems (AMT). Their research is based on Teece conceptualization of long-linked versus intensive technologies. Using their classification scheme, DMS are considered long-linked industrial systems employing hard automation whereas AMT are post-industrial enterprises employing flexible resources.

There are many distinctions between craft shops and DMS. While craft shops employ skilled artisans who use various hand tools, DMS deploy special-purpose machinery operated by unskilled manual laborers. In a craft shop, workers are organized into task-oriented work groups. Work is functionally specialized and used usually a single task is assigned to a worker in a DMS. In a craft shop, performance measures are based on customs and other craftsmen evaluate the work. DMS are controlled through a hierarchical structure. In addition to the above characteristics, DMS are product-oriented, concerned with efficiency and productivity. An information system controls task execution and co-ordinates sequential activities within a DMS.

Adoption of AMT

The study of the adoption and dissemination of technologies is one of the key components of innovation and technological development. Indeed, it is through the adoption of newer, more advanced technologies that industries can increase their production capabilities, improve their productivity, and expand their lines of new products and services. Adoption of new technologies is a key element to a firm's success. Therefore, this report outlines the extent to which establishments in the Canadian manufacturing sector use advanced technologies. It investigates the extent to which advanced technology is used - both at the individual technology level and at the functional technology group level, where functional group refers to collections of technologies that serve a common purpose. Not all firms adopt advanced technologies because of the costs associated with their adoption. Adoption occurs when the benefits from adopting the new technology outweigh the costs.

Michael Proter (1986) argues that there are two roads to competitive advantage: lower cost or product differentiation. Developing countries are able to again a foothold in international markets for their industries by translating their relatively cheap labor costs into lower-cost goods. For industrialized countries such as the US, where the cost of labor is relatively high, the introduction of such cost competition has undercut many traditional industries, and forced many firms to slash costs in order to remain competitive. In the long run, such strategies cannot pay off for industrialized countries. The competitive advantage of low cost production will always lie with less developed countries.

According to Porter (1986), the road to competitive advantage for industrialized countries that wishes to keep wages relatively high must be product differentiation. Firms must become able to produce goods that cannot easily be transferred to lower cost locations, allowing them to maintain reasonable operating margins while paying higher wages.

Global competition continues to drive the adoption of AMT. To date, the literature AMT on the introduction of AMT new technology is based primarily on case studies and expert opinions.

Global and domestic firms have different objectives for adopting AMT as a means to effectively compete in their respective markets. Whatever the objectives may be, the adoption of any new technology involves uncertainty about achieving the objectives.

In addition to the inherent human resistance to change and innovation, at least two other types of uncertainty are present when adopting manufacturing innovations: technological uncertainty (whether the adoption of the technology will be profitable): and strategic uncertainty (how the decisions of competing firms will be adversely impact a decision to adopt a new technology). Generally, the effects of technological uncertainty can be reduced by research and testing. In contrast, strategic uncertainty is more difficult and problematic to evaluate, frequently relying on speculative efforts to anticipate the decisions of rival firms. Adopting AMT involves both types of uncertainty.

Baldwin (1995) suggests that firms, which export a large portion of their production, adopt a greater variety of AMT than domestic competitors. He cites the reorientation to exporting by the Japanese as their motivation to adopt AMT. If this is the case, US exporting firms would be expected to adopt a greater variety of AMT than non-exporting firms.

The adoption of AMT involves major investment and a high degree of uncertainty and, hence, warrants considerable attention within manufacturing firms at the strategic level. As a result, issues involving selection and justification procedures assume greater importance. Baldwin further states that companies can attain significant competitive advantages through AMT such as flexible manufacturing systems (FMSs), computer-aided design and robotic systems. He also observes that many companies are reluctant to install these technologies because:

- Those which have frequently do not reap the advantages these technologies can offer.
- There are difficulties in implementing the expensive, complex systems.
- There are insufficient internal skills.
- There is a multiplicity of implementation paths.
- AMT involves incremental skill building
- AMT requires different support infrastructure

Implementing AMT is one of the lengthiest, expensive and complex tasks a firm can undertake.

An issue of importance in this strategic decision is whether the process should be transferred without modification or adapted in some way for transfer, cloning a manufacturing process can maintain commonality throughout a global network of operations, and avoid re-engineering costs. Cloning, however, requires 'robustness' to the recipient's local conditions and may not allow the exploitation of benefits from local factors of production. The alternative is to adapt some aspect or aspects of the manufacturing process, through adoption; the transferor (the home) can take advantage of local characteristics and can facilitate the transfer process. Manufacturing process which receives significant inputs from the local economy, such as skilled labor, repairmen, reliable power, spare parts, industrial materials processed according to exacting specifications and so on, are less appropriate to the less-developed areas than those which do not have such requirements. Vernon concluded that firms would either vertically integrate to overcome such supply problems, or choose products which were more transferable, such as those which did not require the local conditions mentioned above.

Taking a largely economic perspective, he identified the variables affecting technology choice for transfer as: market size and growth, labor and capital costs, range of technology available and prospect of technology obsolescence. In these examples, the author has listed both host-dependent and host-independent issues. To attempt a comprehensive assessment of the factors coming to bear during the transfer it is necessary to differentiate between those characteristics of the process that pertain to fit with host conditions (such as labor costs and market) and those that affect the activity of transferring the process (such as dependence on difficult-to-transfer knowledge).

Literature on 'technology appropriateness typically suggests factors influencing adaptation to improve fit with host characteristics and socio-economic objectives. Pakes (1984), for example, focuses on labor substitution, identifying appropriate technology as that which maximizes output and employment simultaneously. This approach to appropriateness is too narrow to be of much value to transfer practitioners assessing a manufacturing process or choosing a location. A wide view of appropriateness should draw in all the factors that have some impact on the operation of a manufacturing process in its new location.

Implicitly so far, adoption and adaptation of manufacturing philosophies by less developed countries (LDCs) provides useful material for the search for factors adaptation. The simplicity of Japanese techniques and their low capital investment requirement makes them ideal for LDCs, requiring only training. However, they do mention caveats, for example: workers in LDCs often lack inherent housekeeping tendencies, capability in tool making and quality measurement, and individuals for troubleshooting. More adaptation influencing factors are suggested by authors in LDCs, namely, employee and supplier participation (cultural and infrastructure), educational level, labor costs, unionization, and firms' size.

Factual evidence of the factors affecting manufacturing process adaptation is uncommon, but a number of empirical studies have identified a lack of host managerial know-how, lack of infrastructure, poor IPR protection, governmental requirements and commercial habits as barriers to transfer. Different manufacturing processes will be affected by these factors to a greater or lesser extent. The sensitivity of a process to any of these factors gives an indication of its robustness. If a process was not, for example, climate specific, then it could be considered robust to climate. In this case it would be appropriate, regardless of differences between the home and host climates. The more general case would be where although the process was affected by climate, the home and host climates were sufficiently similar so that the manufacturing process was appropriate for that host.

Adoption of new technologies is a key element to a firm's success. Therefore, most of the researchers outline the extent to which establishments in the Canadian manufacturing sector use advanced technologies. It investigates the extent to which advanced technology is being used-both at the individual technology level and at the functional technology group level, where functional group refers to collections of technologies that serve a common purpose. Not all firms adopt AMT because of the costs associated with their adoption. Adoption occurs when the benefits from adopting the new technology outweighs the costs. Adoption rates, alone, are insufficient for attempting to understand the complex nature of technological change. Thus, some surveys investigate the benefits and effects that manufacturing establishments receive as the results of adopting AMT.

Moreover, some manufacturers hold the view that the adoption of AMT involves a high level of investment, and its payback period is usually longer than that traditionally required by business enterprises.

The top five obstacles to more rapid adoption of AMT, as stated by these executives, are:

- Lack of necessary funding
- Lack of in-house technical expertise
- Failure of top management to grasp the benefits of AMT
- Inadequate planning or lack of vision
- Inadequate cost-justification methods

BENEFITS OF ADVANCED MANUFACTURING TECHNOLOGY

The literature identifies a variety of technical and strategic factors that include AMT adoption: reduced product development time, labor costs savings, material costs savings, a need to remain competitive, tax incentives, financing availability, a need for product change flexibility, environmental, safety or health concerns, increased profitability or plant performance and customer requirements. These factors have a broad, strategic impact on the firm and affect virtually every major element of a firm's operating environment.

AMT represents a wide variety of modern computer-based systems devoted to the improvement of manufacturing operations and thereby enhancement of the firm competitiveness. AMT, in its varying forms, has been credited with the potential to bestow, among other things, earlier entrance to market, faster responses to changing customer needs, and higher quality products with improved consistency and reliability, However, results of several empirical studies indicate that, while most firms achieve some benefits, many of them are not fully exploiting their AMTs touted capabilities.

Since the technical capabilities of AMT are well proven, failure to achieve the potential benefits has been attributed to infrastructure problems such as inadequate organizational planning and preparation for the adoption of the AMT or defines a life-cycle implementation process which consists of the following three phases: pre-installation (planning and justification), installation and commissioning (monitoring and evaluation). Implementation is typically viewed as a combination of the actions in the installation and commissioning and post-commissioning phases.

DISADVANTAGES OF AMT

Companies have identified some of the issues and problems arising from implementing an AMT project and discuss the following issues, namely scope of AMT projects, simulation modeling, cell design, cell operational logistics, and labor issues in AMT.

Traditionally there has been little awareness of the need to link technology issues with human resource management issues. Technologies have been deployed to improve firm performance independent of programs and practices that involve workers in decision-making. Firms implement new technologies to improve their competitiveness. Reductions in labor costs are made possible by new automated machines and equipment which take over decisions about production and perform at accelerated rates operations formerly performed by humans. Faster performing technologies permit increased production using the same fewer, or less skilled employees.

The impediments that were investigated by Baldwin can be divided into five groups. The first includes a set of general cost-related problems associated with AMT adoption, including the cost of capital, the cost of technology acquisition, the cost of related equipment acquisition, the cost of related software development, and increased maintenance expenses.

Four other areas were also identified-impediments that arise from government policy (Institutional-related problems), from labor market imperfections, from internal organization problems and from imperfections in the market for information. Each of these also increases the costs of adopting AMT-but the causes are somewhat more narrowly focused than the general cost-related items that are included in the first category.

Labor-related problems include difficulties that arise because new technologies and innovation generally need higher skill levels. In the face of these needs, a firm may encounter impediments to adoption if there is a shortage of skills available on the market, or if it faces training difficulties in overcoming deficiencies, or if its labor contracts act to constrain its ability to substitute labor across tasks.

Organizational problems are those associated with difficulties in implementing the types of internal change in a firm that are required for the adoption of AMTs. The first of these is the difficulty in introducing important changes to the organization. For example, the introduction of computer-aided design may require new structure that link engineering development with the production department so that the advantages of concurrent engineering practices can be fully exploited. Other organizational problems stem from a poor attitude of senior management towards new technologies, or worker resistance.

Institution-related problems are those associated with tax regimes (both the R&D tax credit and capital cost allowances) and with government regulations and standards. Information-related problems arise if markets for knowledge are imperfect. They include lack of scientific and technical information, technological services, and technical supports from vendors. The percentage of plant managers who reported that these problems impeded the adoption of AMT according to the Baldwin (1995) survey.

Table. Problems Impeded the Adoption of AMT

Types of Impediment	Percentage
Cost-related	79.1
Institution-related	15.5
Labor-related	37.2
Organization-related	34.1
Information-related	24.9

Role of AMT in Sri Lanka

Almost all developing countries that were British colonies for a considerable length of time inherited the British Education system. During the early years of the colonial period, British investors set up most of the sizable businesses in these countries. The managerial personnel, including Production Managers, for these enterprise were generally brought from the UK. Singapore and Sri Lanka were British colonies for nearly one hundred and fifty years. Both of these countries inherited their education system and practice almost entirely from the British systems.

During the colonial period, business activity in Sri Lanka was directed towards the plantation sector introduced to the economy by the British. In order to facilitate the investment of British capital, plantation joint stock companies were introduced in the middle of the nineteenth century. The Sri Lankan government has found that companies gain and sustain international competitive advantage through improvement, innovation and upgrading. Research in the early 1990s showed that many Sri Lankan companies were failing to compete effectively. Productivity in Sri Lankan firms had been relatively low by other countries standards.

Various attributes are addressed and used by researchers for procedures involving adoption, classification, selection, and justification of AMT. In this review, a comprehensive list of studies have been identified and classified under many categories. The literature also observes common difficulties in implementing AMT such as lack of technical skills, managerial problems, lack of confidence to implement automated systems, lack of clear-cut policy direction towards automation, resistance to adapt automation and systematic evaluation methods. Economic issues alone are inadequate to justify new manufacturing systems because traditional evaluation methods are inadequate for the purpose. Non economic benefits could not be included in the justification procedure, which offer a large number of intangible benefits. The problem lies not in the level of technology, but rather with its implementation. It is important to note that, instead of rushing to invest in AMT, a manufacturing company must reassess its direction, strengths and weaknesses, and develop a strategy for successful implementation accordingly. The entire literature review and classification scheme suggested have brought several elements to the fore. These can be summarized as follows:

- AMT adoption factors and adaptation and implementation problem by focusing on issues of specific interest.
- AMT involves a set of quantifiable and non-quantifiable attributes.
- AMT definition has changed from time to time. It starts from broad definition and ends up with a much computer related field.
- AMT includes many issues, especially human issue and mechanical issues.
- Advantages and disadvantages related to AMT studies are common in the literature.

ADVANCED MANUFACTURING

Advanced manufacturing is the use of innovative technology to improve products or processes.

USE OF TECHNOLOGY TO IMPROVE PRODUCTS AND PROCESSES

One of the most widely used definitions of advanced manufacturing involves the use of technology to improve products and/or processes, with the relevant technology being described as "advanced," "innovative," or "cutting edge." For example, one organization defines advanced manufacturing as industries that "increasingly integrate new innovative technologies in both products and processes. The rate of technology adoption and the ability to use that technology to remain competitive and add value define the advanced manufacturing sector." Another author states: "Advanced manufacturing centers upon improving the performance of US industry through the innovative application of technologies, processes and methods to product design and production." Finally, a recent survey of advanced manufacturing definitions by theWhite House states: "A concise definition of advanced manufacturing offered by some is manufacturing that entails rapid transfer of science and technology (S&T) into manufacturing products and processes." (PCAST, April 2010.)

Products

One organization characterizes advanced manufacturing products as follows:

- Products with high levels of design
- Technologically complex products
- Innovative products
- Reliable, affordable, and available products
- Newer, better, more exciting products
- Products that solve a variety of society's problems

Process Technologies

The manufacturing process technologies described in definitions of advanced manufacturing include:

- Computer technologies (e.g., CAD, CAE, CAM) (Paul Fowler, NACFAM; UK Manufacturing Advisory Service Southeast; C.B. Adams, St. Louis, OECD)
- High Performance Computing (HPC) for modeling, simulation and analysis (Council on Competitiveness)
- High Precision technologies (Paul Fowler, NACFAM)
- Information technologies (Paul Fowler, NACFAM)
- Advanced robotics and other intelligent production systems (US Department of Labor, ETA; C.B. Adams, St. Louis)
- Automation (UK Manufacturing Advisory Service Southeast; C.B. Adams, St. Louis)
- Control systems to monitor processes (UK Manufacturing Advisory Service Southeast)
- Sustainable and green processes and technologies (US Department of Labor)
- New industrial platform technologies (e.g., composite materials) (UK Manufacturing Advisory Service Southeast)
- Ability to custom manufacture (PCAST; Paul Fowler, NACFAM; Grow Oklahoma Campaign)
- Ability to manufacture high or low volume (scalability) (PCAST; Paul Fowler, NACFAM; Grow Oklahoma Campaign)
- High rate of manufacturing

USE OF BUSINESS/MANAGEMENT METHODOLOGIES

A number of organizations also included business or management methodologies in their definition of advanced manufacturing. For example, one organization defines "advanced manufacturing as the insertion of new technology, improved processes, and management methods to improve the manufacturing of products." (National Defense University, 2002, as reported in PCAST) Another organization lists advanced manufacturing as "encompassing lean production techniques, enhanced supply chain integration, and technology assimilation." In fact, the Wikipedia definition of "advanced manufacturing" is "advanced planning and scheduling" described as "a manufacturing management process by which raw materials and production capacity are optimally allocated to meet demand." Overall, the following business or management methodologies were listed as being a part of advanced manufacturing:

- Quality controls (US Department of Labor)
- Lean production technologies
- Supply chain integration

- Advanced Planning and Scheduling (Wikipedia, online, accessed June 2010)

OTHER DEFINITIONS

There are also definitions of "advanced manufacturing" that are used by one or few sources.

Traditional vs. Advanced Manufacturers

In one report, the distinction between traditional sectors of manufacturing (listed as auto, steel) and others (listed as aerospace, medical device, pharmaceutical) is the basis for a definition of advanced manufacturing, with the characteristics of the two differing in terms of volume and scale economies, labor and skill content, and the depth and diversity of the network surrounding the industry. (New England Council and Deloitte, as referenced in PCAST document)

Successful Manufacturers

Other sources define advanced manufacturers as those that "succeed" in today's competitive environment. One source states that: "What differentiates certain companies is a unique ability to create a competitive advantage in this environment. These manufacturers think and do faster and, by definition, these advantages make them advanced." (Industrial College of the Armed Forces) The White House survey lists some experts as defining advanced manufacturing "solely by advances that led to decreased cost or increased productivity." (PCAST)

Research and Development

One organization listed "aggressive research and development" as being part of the definition of advanced manufacturing. (Purdue University) Although research and development was not explicitly included in most definitions, the innovative technologies listed by many are most likely the result of extensive research and development.

DYNAMIC, CONSTANTLY CHANGING

Finally, several sources pointed out that any definition of advanced manufacturing will need to change with the changing times, and that the definition will vary for different companies and different industries. The White House survey states: "Most discussants agree that an appropriate advanced manufacturing definition should be dynamic in nature and be treated as more of a benchmark. That is, there is a constant iteration of improving manufacturing frontiers...Therefore, what is classified as "frontier" is constantly changing, and, likewise, advanced manufacturing is constantly changing." (PCAST) Another source stated that: "Advanced manufacturing is like a chameleon. It changes in response to the needs of whichever company has incorporated it into its manufacturing process." (St. Louis, C.B. Adams) A final expert is quoted as saying: "Advanced manufacturing, by its very nature, defies definition, because it is going to be different for the chemical industry than it is for the metal fabrication industry and any other industry." (Tom White, as quoted by C.B. Adams)

Conclusion

The term "advanced manufacturing" encompasses many of the developments in the manufacturing field during the late 20th and early 21st centuries, including high tech products and processes and lean, green, and flexible manufacturing, among others. No one definition captures everything said about advanced manufacturing, although the majority of definitions found on the web include the use of innovative technology to improve products and/or processes, and many also include the use of new business/ management methodologies. Accordingly, the definition that probably comes closest to being comprehensive is that given by Paul Fowler of the National Association of Advanced Manufacturing (NACFAM), celebrating its 20th anniversary this year:

"The Advanced Manufacturing entity makes extensive use of computer, high precision, and information technologies integrated with a high performance workforce in a production system capable of furnishing a heterogeneous mix of products in small or large volumes with both the efficiency of mass production and the flexibility of custom manufacturing in order to respond quickly to customer demands." (Quoted in PCAST)

ADVANCED MANUFACTURING TECHNOLOGY AND COMPUTER INTEGRATED MANUFACTURING

AMT are the main technical components of Computer Integrated Manufacturing (CIM) systems. Although a CIM system contains many AMT such as Computer Aided Manufacturing (CAM), Computer Aided Design (CAD) and Computer Aided Process Planning (CAPP), it is more than a group of advanced and automated technologies. CIM is the term used to describe the modern approach of manufacturing (Haywood, 1990).

The main feature of CIM is the total integration of all manufacturing functions, including design, engineering, planning, control, fabrication, and assembly etc. through the use of computers. So CIM is a comprehensive measure of computerised integration and information sharing in a manufacturing system. According to the CIM wheel model by Society of Manufacturing Engineer (SME), there are one business and four technical components of a CIM system (Goetsch 1990). The four technical components are planning and controlling, information resources management, product and process definition, and factory automation. The four components and relevant AMTs involved are described below (Groover, 1987; Goetsch, 1990; Singh, 1996; Kotha and Swamidass, 1998).

The planing and controlling component includes such element as planning/scheduling and controlling of facilities, materials, tools and shop floor activities. Hardware and software are available to automate each of the elements. MRP, as well as Manufacturing Resources Requirement (MRP II), is an important concept with a direct relationship to CIM. MRP involves using the bill of materials, production schedule, and inventory record to produce a comprehensive, detailed schedule of the raw materials and components needed for a job (Chase and Aquilano, 1997). As other manufacturing technologies have evolved from automation to integration, MRP has also developed. The new version of MRP, known as MRP II, goes beyond determining materials requirement to also encompassing financial tracking and accounting. MRP II is particularly well suited to the integrated approach represented by CIM.

Information recourses management is the nucleus of CIM. Information, updated continually and shared instantaneously, is what CIM is all about. One of the major goals of this nucleus is to overcome the barriers that prevent the complete sharing of information among all other CIM components. The AMTs used for this purpose include Shared Databases (Shared DB), Wide Area Network (WAN), and Local Area Network (LAN). Each of these represents different levels of information integration and sharing. In addition to these standalone and islands of automation technologies, the term CIM is also a comprehensive measure of computerized information sharing.

The product and process definition component of the CIM wheel contains three elements: design, analysis and simulation, and documentation. This is the component where products and process are designed, engineered, tested through simulation, and documented through drawing specifications and other tools. Standalone technologies include Computer Aided Design (CAD) which can be used to automate the drawing and analysis process and Computer aided Engineering (CAE) which can automate the simulation and analysis process. The islands of automation in this component is Computer Aided Process Planning (CAPP) which aims to link design, engineering and manufacturing processes by converting design parameters into processing codes.

The factory automation component contains those elements that are associated with fabricating and assembling products, for example, material handling, assembling, inspecting and testing, and materials processing (i.e., fabricating). The AMT technologies suited to these elements include Numerical Control/Computer/Direct (NC/CNC/DNC), Computer-aided inspection/ testing/tracking (CAI/T/T), Computer-Aided Manufacturing/FMC/FAS (FMS/ FAS), Automated parts loading/unloading (APL/U), Automated tool changes (ATC), Robotics (Robot), Automated Storage/Retrieval Systems (AS/AR), and

Automated Guided Vehicles (AGV). AMTs are also classified according to the degree of automation and integration. Bessant and Haywood (1988) suggested four levels of integration. They are standalone, islands of automation, archipelago of automation (i.e., partial integrated) and the full integrated systems.

Standalone AMT refers to single machines or equipment that are not directly connected with other machines or systems by computers. NC machine is a typical example of standalone AMT in fabrication and a single CAD is a standalone AMT in design process.

An island of automation refers to a special group of automated machines that work together but have no direct communication with other machines and systems outside their group. FMS is a typical island of automation is manufacturing. Island of automation exists also in design, engineering and process planning processes.

Integration refers to the connection of at least two different functions by computer. For example A CAPP can link design and engineering processes by converting design parameters into manufacturing plan and codes. MRP II system can link design, manufacturing and finance functions to dynamically update the changes of raw materials or components. Integration varies from partially integrated to full integrated.

RESEARCH METHOD AND EMPIRICAL DATA

The survey

The research reported in this paper is based on the data from the International Manufacturing Strategy Survey IMSS. IMSS was initiated by London Business School and Chalmers University of Tech¬nology and is being co-ordinated by Instituto de Empresa in Spain. A worldwide researcher network in more than 20 countries carried out the survey.

The questionnaire was first designed by Swedish team in 1980s and modified for the first international survey in 1992-1993 period. It was modified again for the second round of survey based on the experiences from 20 countries in the first round of survey. The questionnaire was discussed in a workshop participated by IMSS researchers. For details of the IMSS project, please refer to the book by Lindberg, Voss, and Blackmon (1998).

The questionnaires were sent to companies in individual countries, separately, in the period from 1997 to 1998. The methods of data collection vary from country to country. In some countries, post survey was used, while in others, on-site interview was employed. All the data were sent to the co-ordinator in Spain and then distributed to all participants. As a participant of the IMSS project in Denmark, Norway, China and Hong Kong, the author has the right to use the data for the purpose of teaching and research. Data from 18 countries were available when this research was conducted. The participating countries and the sample size (in parentheses) are as follows: Argentina (31), Brazil (27), China (30), Denmark (27), Finland (14), Hungary (38), Italy (71), Japan (29), México (29), Netherlands (29), New Zealand (32), Norway (13), Perú (8), South Korea (50), Spain (33), Sweden (27), the UK (24), and the US (41). Altogether the sample size is 556.

The sizes of the 556 sampled companies vary. Since the standard for small, medium and large size companies vary from country to country and this project includes 18 countries, the sampled companies were divided into five classes according to their size: <100 (14%), 101-500 (40%), 505-1500 (26%), 1001-3500 (11%), and >3000 (6%) employees. Missing data occupies about 3%.

The companies joined the survey are in the International Standard Industry Classification (ISIC) 38 group. Example products in ISIC 38 industry include metal products, machinery, electrical machinery apparatus, appliances and suppliers, transport equipment, professional and scientific measuring and controlling equipment, and photographic and optical goods. Companies in ISIC 38 industries adopted more AMT than others (Waterson et al, 1997).

AMT uses and payoffs

The above mentioned ques¬tionnaire covers 300 variables about strategy, practice and performance. The research reported here focuses on the uses, payoffs and contributions of AMT. The AMTs measured in this survey. The degree of use, the relative payoff, and the adoption of the AMT within next three years are measured. The measures are all at 1-5 scale, which ranges from 1: None, 2: Little, 3: Some, 4: Much to 5: Very High.

Performance

Performance improvement is measured by the percentage of changes in the past three years. It covers twenty four items as shown in table. Factor analysis was conducted with a view of reducing data and identifying factors of performance. The result is that the twenty-four items converged into one performance factor (PF). Data analysis will use this performance factor in stead of individual items.

RESULTS AND DISCUSSIONS

The results from this research will be presented in the following parts, corresponding to the current uses of AMT, the expected uses of AMT in three years, and the relationship between AMT uses and performance.

The current uses of AMT

Individual AMT

Based on the means of current uses, the sixteen AMTs are ranked as shown in the column of "AMT uses in 1998" in table. According to the rank by means, the top five AMTs are CAD, MRP, LAN, CNC, and shared Database. These AMTs are more commonly used than others. For example, 80% of sampled companies used CAD from some to a lot. The six AMTs at the bottom are AGV, AS/AR, CIM, robots, and automated tool change. These AMTs are rarely used by the sampled companies. For example, 18% companies used AGV from some to a lot while 76% did not used have AGV at all.

Automation and Integration level

With regard to integration levels, standalone technologies are more commonly used. Among the top five commonly used AMTs, four are recognised as standalone. They are CAD, CNC, MRP, and LAN. The second popular technologies are those in the stage of islands of automation, such as FMS. Partially and fully integrated technologies such as CAPP, WAN and automated material handling and transportation technologies such as AGV etc. are least used. This indicates a sequence of adopting AMT, i.e., from simple, standalone, islands of automation to integration. This is useful for those companies, which did not have any AMT at all and start to think about where to start. Companies can start with CAD in engineering, CNC in manufacturing, and MRP in administration. According to this pattern, it can be found that the most commonly used technologies are relatively simple and standalone technologies, while the less commonly used AMTs are complicated and advanced. This also indicates that fully integrated and automated factories are still not a reality. Taking the measure of "CIM" as an example, only 5 of the sampled companies claimed they use CIM at a full scale while nearly half of the companies did not implement any CIM at all.

CIM components

The commonly used AMTs (i.e., the top five) are found in all the four CIM components as shown in table. However, based on the average degrees of uses in the four areas, the ranks are: planning and control, design and engineering, information resources management, and factory automation. The lower average score of AMTs used in factory automation is mainly due to the low uses of automated material handling and transportation technologies. The more uses of AMT in design and information management indicates that companies pay more attention to product varieties and the speed of product development, which may reflect the tendency of shorter product life cycle and technology development in the international market. In each of the CIM component, there seems also to be an adopting sequence from simple, standalone, islands of automation, to integrated systems.

Expected use in three years

Individual AMTS

Compared with AMT uses in1998 expected AMT uses in three year will all increase. The differences between 2001 and 1998 are all significant at a level of 0.001. This is promising for companies both producing and using AMTs. The top 6 AMTs will be adopted by companies within the next three years are CAD, LAN, shared Database, MRP, and MRP II. The top fives according to their ranks are the same as they are in 1998. However, within the top five, LAN and shared Database increased much more than others. They become number two and number three. Both for the moment and within the next three years, CAD is ranked very high, which indicates that product design and development are emphasised. This is explainable considering the demanding on shorter time of product development by the market nowadays. Those AMTS for handling and transporting materials will still be at the bottom in three years.

Level of integration

Although all increase, the degrees of increase in AMT uses vary. Considering the current uses and changes in three years, AMTs can be divided into three groups.

Group I includes those advanced material handling technologies such as robots, AS/AR, Auto PL/U, Auto TC, AGV. These technologies ranked very low in 1998 and the changes are also rather lower relatively. These technologies are the main components of highly integrated manufacturing system, especially in material handling and transpiration integration. This implies that, in the near future, the materials handling and transporting will still be far from fully computerised integration.

Group II includes those widely used AMT such as MRP, CAD and NC machining tools and LAN. These are typical standalone AMTs. Their uses in 1998 are quite high. However, their increase in three years will relatively be low.

Group III includes those technologies in the stage of islands of automation such as FMS and CAE etc. These technologies are used in the medium degree. However, the increase of their uses in three years will be on the top among the three groups, especially CAPP and shared database.

It is expected that the integration level will be increased in three years. This can be demonstrated by the increase in uses of AMT that connecting more than one function such as shared database and CAPP.

The components of CIM systems

Comparing 1998 and 2001, uses of AMTs in all the four CIM components. For all the four CIM components, the differences between 2001 and 1998 by t-test values are 15, 10, 17 and 17, all significant at the level of 0.001. The differences in information management and design and engineering are bigger than in other two. Since information sharing is the main feature of CIM system, the increase in information management may indicate that in the future, not only individual AMT uses increase, the level of integration will increase as well. Increase in uses of design and engineering reflects the companies will continue to emphasise product development.

In 1998, AMT uses in the four CIM components are significantly different. The paired sample t-tests (significant levels) between the four CIM components are 3.3 (0.001), 7.3 (0.001) and 7.5 (0.001). However, in 2001 the differences will be smaller. The paired sample t-tests (significant levels) are 0.16 (0.87), 1.7 (0.09) and 9.4 (0.001). The numbers of t-tests and significant levels. Except for in factory automation, the degree of AMT uses in other three components are similar. This means that AMT uses in these three components are more evenly used in the three CIM components. However, the AMT uses in factory automation will still be lower. The above evidences indicate that in the near future, although companies will move further towards CIM, full integration in factory automation will still be lagging behind.

AMT AND PERFORMANCE

The relationship between the use of AMT in 1998 and the payoffs and performance improvement will be investigated. Payoffs of AMT are measured by two means. The first is to ask the mangers to subjectively evaluate the payoff of AMT at a 1-5 scale (cf., section 3.2).

The second is to investigate the empirical relationship between the degree of uses of AMT and the improvement of business performance such as cost reduction, quality improvement etc. (cf., section 3.3). The relationship will be discussed for individual AMTs, the four CIM components and the level of integration.

Individual AMTs

The ranks of AMTs in terms of payoff. The five AMTs which are on the top in terms of payoff are CAD, LAN, MRP, CNC and shared database. The five AMTs at the bottom are CIM Computer-Integrated Manufacturing, Robotics, Automated tool changes, AS/RS Automated Storage/Retrieval Systems, and AGV's Automated Guided Vehicles.

The degree of payoffs is correlated with the degree of uses. The correlation coefficients vary from 0.80 to 0.98 and the significant levels are all less than 0.001. It can be concluded that managers believe that the more AMTs are used, the higher will payoffs tend to be. From this finding, it can be proposed that, in order to get more benefits from AMT, companies should invest AMT to great extents.

The finding proposes a simple linear relationship between the uses of AMT and the payoffs. However, based on the correlation and ANOVA test of the relationship between the uses of AMT and the improvement of business performance, several different patterns of relationships are identified, which are discussed below.

The ladle-shape relationship

The relationship between the uses of FMS/C, CIM, MRP and performance improvement is like a ladle shape curve. From none to some degree of AMT uses, performance improvement does not increase and even drops. There seems to be a lead-time between implementation and benefit show up. This indicates a learning curve effect in the process of implementing this type of AMT technologies. Companies should not be frustrated at the slower showing of benefits from these technologies.

The bell-shape relationship

The relationship between the uses of CAD, CAPP, CAE and performance improvement is like a bell shape curve. From none to some degree of AMT uses, performance improvement increases. However, from some to very much degree of AMT uses, performance improvement decreases, which seems to be strange. The explanation may be that, when technologies are implemented to a great extent while organization and human sides are not innovated accordingly, the success rate of AMT tends to be lower. This has been demonstrated by many researchers (Voss 1988; Bessant, 1990, Sun, 1994). This is also true for design technologies (Twigg and Voss 1992). This group of technologies is all used in design and engineering. So it is suggested that when AMT in design are implemented to high extents, organization and human development should also be considered. In other words, AMT uses should be in a balance with the development of human side.

The irregular relationship.

The relationship between performance improvement and the uses of Shared database and Computer Aided Inspection and Testing are up and down, showing an irregular pattern. It is very misleading if the relationship is significant according to simple linear correlation analysis. For example, simple linear correlation reveals a significant correlation between the degree of uses of shared database and the improvement in performance. According to the logic of conventional management research the conclusion can be draw as " the more this technology is used, the higher will the improvement in performance".

No relationship at all

For some AMTs, there is not any relationship at all, neither by correlation nor by ANOVA test. Half of the AMTs belong to this group. However, it should be mentioned that no negative relationship was found, neither. So it does not mean that these AMT are not contributing. They may contribute in some companies, but not in others. The relationship may be rather complicated. Future research may have to look at the methodology on relationship between AMT and performance improvement

CIM components and performance improvement

Individual AMT, whose relationship with performance improvement show certain patterns, are found in all the four CIM complements, for example, CAD in design and engineering, MRP in planning and control, shared database in information management and CAI/T/T in manufacturing. Looking at the average AMT uses of the four CIM components, planning and controlling AMTs have a significant correlation with the improvement of performance, as shown in table. The average uses of AMTs in design and information management are also correlated with performance improvement. However, there is no significant correlation between performance improvement and AMTs uses in fabrication and assembly. Maybe this is due to the low uses and low payoffs of those automated material handling AMTs. Since the integration level in material handling and transportation is still low, contribution is more individually rather than collectively. With the increase in integration level, the contribution of CIM components may be more obvious. The result, on the one hand, verified the CIM wheel model by SME and, on the other, provides another way of investigating AMT uses and performance improvement.

Level of integration and performance improvement

It was found that relationship between performance improvement and most of the standalone and islands of automation show certain patterns (linear, ladle or bell). However, there is not any obvious general pattern about the relationship between performance improvement and those advanced and integrated material handling and transportation AMTs such as AGV, AS/AR, Robots etc.

In summary, the relationship between AMT uses and contribution are much more complication than expected. The implication for methodology is that simple linear correlation bears great danger in drawing conclusion about the relationship of any two variables. Simple correlation may lead to simplistic conclusion such as "the more is an AMT used, the higher will be the improvement of performance". However, in most cases of AMT, such pattern of relationship does not exist. Conclusions from simple linear correlation may be misleading. For example, simple relationship reveals a significant positive correlation between the use of MRP and performance improvement, which implies that the more MRP is implemented, the higher the improvement of performance. However, the real relationship is like a ladle shape, which implies that at the beginning stage of implementing MRP, the performance may not be improved much.

CONCLUSIONS AND IMPLICATIONS AND CONTRIBUTIONS

This research revealed that CAD, MRP, LAN, and CNC machines are the most popular AMT. It seems that there is a sequence of adopting AMT, namely from simple to complicated. Fully integrated and green field CIM seems to be rare. In three years, the uses of CAPP and shared database will significantly increase, which indicate the increase in integration level of manufacturing system. However, the main configuration of manufacturing will be standalone, islands of automation, and limited integration. Fully computerised integration in manufacturing system will unlikely be the main model in the near future.

Regarding the relationship between AMT uses and performance improvement seems to be complicated than expected and reported in references. Simple linear correlation and ANOVA are the conventional method for investigating the relationship between two variables. In this research, both methods and scatter graph were used. A couple of difference patterns of relationship between AMT uses and performance improvement were identified and methodological issues and practical implications are useful.

The contribution of this research include: 1) reveal the extent and patterns of AMT uses, 2) reveal the tendency of AMT uses in the next three years, and 3) reveal the relationship between AMT uses and performance.

Practical Implications

The results on the pattern and tendency of AMT uses and contribution will be of reference to companies either using or producing AMTs.

It seems that there is sequence from simple, standalone, islands of automation to integrated AMTs. This may be a guide for companies that have not implemented AMT yet. Companies normally started with CAD in design and engineering, MRP in planning and control, LAN in information sharing and management, and NC in factory automation. AMT will be mainly in in the stages of standalone and island of automation. Fully integrated manufacturing systems are not likely in place in the near future. The adoption process is a step by step and evolutionary. Neither vendors nor governmental organisations should promote fully integrated and green-field manufacturing systems as in the 1980s.

Regarding the relationship between AMT uses and contribution, managers' subjective evaluation seems to support a linear relationship between AMT uses and performance improvement. However, statistic analysis reveals several difference patterns of relationship between AMT uses and performance improvement. First, managers should be noticed that the relationship could be much complicated than expected. Second, for some AMT technologies, there is a learning curve effects. It really takes time before the benefits show up. Finally, AMT uses should be accompanies by organisational and human development.

Limitations and future research

Among the sampled companies, the percentage of small companies is rather small (cf., section 3). This implies that the results of this research should be limited mainly to large companies. Future research can be conducted either with only small companies or to compare the small and large companies.

Since the data sample of each country are not large enough to represent each country, this research only focused on identifying the overall pattern and tendency of AMT uses and contribution. In future research, the differences in each country should be investigated so that the relationship between national context and AMT uses can be identified and companies in different countries can learn from each other. This information will also be of reference for those companies that would like to establish companies or joint ventures in other countries/regions.

This research has revealed that the relationship between AMT uses and performance improvement is rather complicated. This suggests that future research should not only reply on simple linear correlation to investigate the relationship between AMT uses and contributions. This research also revealed that the managers' subjective evaluation of AMT payoffs were different from that revealed by statistical analysis. Future research should provide a list of AMT payoffs and ask managers to select.

2

Integrated Manufactured Goods and Process Design

INTRODUCTION

Information age manufacturing begins with information age design. Designing both a product and the processes by which it is produced involves understanding what the product is to do and how the product will do those things, converting the requirements for the product's behavior into engineering specifications, and producing plans that marshal the materials, equipment, and people needed to make and deliver the products. Even apart from the pressures for shorter time to market for new products that stress current design paradigms, new design challenges are generated by new product trends (e.g., shrinking feature size, decreasing tolerances, or a growing numbers of parts) and by new manufacturing processes (that designers must learn to exploit).

As a result, the amount of knowledge and data relevant to product and process design is rapidly becoming more than a single individual can comprehend. A further complicating factor is that an integrated product and process design (IPPD) effort must usually be coordinated among a number of engineering teams with different specialities and from different companies, since it is rare for a single company to have all the skills, technologies, and financial resources to design in-house all of the components needed. Managing this coordination task represents a major opportunity for information technology (IT) to have a positive impact. A comprehensive IPPD system would include integration of performance specifications, conceptual design, detailed design, manufacture, and assembly, together with the ability to simulate actual use, field repair, upgrade, and disposal. Such a comprehensive system will not be possible for many years.

However, an important first step toward this vision can be achieved by joining detailed design with manufacturing and assembly. To accomplish this requires a new level of information structuring and integration. Feature-based design is the best way currently known to capture and integrate the necessary information that links the geometry of parts with their functions, fabrication, and assembly. Present computer-aided design (CAD) systems support creation of geometry only. Apart from stress and thermal analyses of parts and certain types of kinematic analyses, most design analyses must be done manually because there is no way to obtain the necessary information from the circles and lines stored in the CAD system.

An IPPD system that realizes this first step toward a more comprehensive system will consist of three elements: a database, a set of algorithms, and user interfaces. The database will be structured to capture the information about the design in the form of geometry plus features (places of interest on each part, together with information on what role they play in the product's function and how to make and assemble the features in relation to each other). The algorithms will take the information they need from this database to simulate function, determine optimal assembly sequences, estimate fabrication or assembly cost, or perform design-for-assembly analyses, for example. The user interfaces will make it easier for the designer to create a design using features and apply the algorithms to study and perfect the design.

Prototype software that does some of these things exists now. This software can be a basis on which to build a new kind of computer-assisted design that integrates technical, business, and economic issues relevant to design. It can support analyses of cost and function, as well as the study of families of products that share parts or subassemblies. Such software has been demonstrated for the design of certain complex electro-mechanical items.

Another IT-enabled connection between design and manufacturing is the use of stereolithography as a visualization aid for designers and as the basis for rapid generation of prototype molds and dies for the production of mechanical parts. In an experiment conducted by a major automobile manufacturer, vendor bids based on a drawing and a stereolithographed model were lower than bids based on a drawing alone; this result was explained by the fabricator's greater ability to visualize the complexities of the item in question and thus to more accurately determine the costs of its fabrication.

Once a core design system is in trial use, it can be extended up and down the design process to include concept design and field use considerations. The information hooks in the data structures will be there to integrate with functional simulations, repair environments, and other aspects of product design. This evolution of the design system will be fueled by experience with its core implementation; feedback from users will determine what new capabilities it needs.

CONCURRENT DEVELOPMENT OF PRODUCTS AND PROCESSES

Concurrent development of products and processes refers to the simultaneous development of the deliverable product and all of the processes necessary to make the product (development processes) and to make that product work (deliverable processes). These processes can significantly influence both the acquisition and life-cycle cost of the product. Process examples include the manufacturing processes needed to fabricate the product, the logistics support processes needed to support the product, or, for a data collection system, the process to collect and disseminate the information gathered. Emphasizing the design of these processes at the same time the product is being designed ensures that the product design does not drive an unnecessarily costly, complicated, or unworkable supporting process when the product is actually produced and fielded. Not developing the processes concurrently with the product results in utilizing an inefficient manufacturing and support process or causing a redesign of the product, which could potentially wipe out any other cost reductions achieved through the application of other IPPD principles.

From an engineering viewpoint, concurrent development of products and processes to satisfy user needs is known as systems engineering. In IPPD, the systems engineering approach to designing a product is expanded to include all stakeholders -- those developing not only the product but all product-related processes as well (e.g., business processes such as financial, contracting, etc.). Multidisciplinary teamwork and an emphasis on real-time and open communication are key to accomplishing this concurrent development. Multidisciplinary teamwork is implemented in an IPPD environment usually through the use of IPTs. Members of an IPT are empowered to make decisions for their respective organizations and keep them informed of the product and process decisions. An enhanced communication environment, where all program information is in a format available to all stakeholders in real time, is of primary importance to the effectiveness of the IPTs.

Early and Continuous Life-Cycle Planning

Early and continuous life-cycle planning is accomplished by having stakeholders, representing all aspects of a product's life-cycle, as part of the multidisciplinary teams. Early life-cycle planning with customers, functional representatives, and suppliers lays a solid foundation for the various phases of a product and its processes. Key program activities and events should be defined so that progress toward achievement of cost-effective targets can be tracked, resources can be applied, and the impact of problems, resource constraints, and requirements changes can be better understood and managed. Early emphasis on life-cycle planning ensures the delivery of a system that will be functional, affordable, and supportable thoughout a product's life cycle.

Proactive Identification and Management of Risk

IPPD is not a "design now-test later" approach to product and process development. Proactive identification and management of risk is accomplished in many ways in the IPPD environment. By using the multidisciplinary teamwork approach, designers, manufacturers, testers and customers work together to ensure that the product satisfies customer needs. DoD endorses a risk management concept that is forward-looking, structured, informative, and continuous. The key to successful risk management is early planning and aggressive execution. IPPD is key to an organized, comprehensive, and iterative approach for identifying and analyzing cost, technical, and schedule risks and instituting risk-handling options to control critical risk areas. IPTs develop technical and business performance measurement plans with appropriate metrics to monitor the effectiveness and degree of anticipated and actual achievement of technical and business parameters. Modeling and simulation tools are used to simulate, test, and evaluate the product prior to starting production. Robust design methods are used to minimize problems in manufacturing and operations. Event-driven scheduling is used to integrate all development tasks and to ensure that a task is not started until all prerequisite tasks are complete.

Maximum Flexibility for Optimization and Use of Contractor Approaches

There are many ways to accomplish IPPD. IPPD is a management approach, not a specific set of steps to be followed. The Government acquisition community recognizes that it must allow contractors the flexibility to use innovative, streamlined best practices when applicable throughout the program. Thus, it cannot specify specific steps for the contractor to follow. The DoD leadership's recent instructions that acquisitions will now be performance-driven, not process-driven, help in maximizing flexibility for the optimization and use of contractor approaches. These instructions allow the contractor more latitude in developing bid proposals and conducting their processes. For example, DoD's efforts to reduce the use of military specifications and standards allows contractors to adapt their fabrication processes and management techniques for optimal use on the product being developed.

APPLICATION TO MECHANICAL DESIGN

Mechanical design poses myriad different problems, and the extent to which the electronic design paradigm can be applied to mechanical design is a matter of some debate. Digital logic can be considered as a special class of product whose design and fabrication problems have proven amenable (with the expenditure of considerable R&D resources) to the application of information technology. By contrast, mechanical items represent a wholly different class of products for which there is currently no formal representation of function and there is no direct algorithm-dominated way for reducing a functional description to physical design; whether this is fundamentally true or merely a limitation on current knowledge is as yet unknown. Most importantly, the science and engineering underlying models of mechanical products and the processes to manufacture them are not nearly as well understood as those for electronic products. Mechanical designs are characterized by complex and large multimedia energy interactions between a limited number of elements and by changes in element behavior over time. The information needed to describe multiple behaviors of mechanical systems is difficult to express in a single format or language, and there are few tools for representing or testing mechanical designs that are comparable in power to those available for the analogous electronic design task. A similar point applies to the modeling of multiple and simultaneous high-level energy interactions. Compared to the problems encountered by the VLSI circuit designer, the problems that challenge the mechanical designer, described in Box 3.2, suggest important research questions for better IT support of the design effort. These challenges are the subject of the remainder of this chapter.

Needs And Research For Mechanical Design

It is not clear that the paradigm of electronic design can be applied to mechanical design, the area that the committee believes poses the greatest need today. Nevertheless, the successes of today's electronic design paradigm suggest research areas to improve mechanical design.

The above three steps may have to be repeated until a functional and manufacturable design in achieved.

Specifically, an important goal is the development of appropriate abstractions at every stage of the design process and tools that manipulate these abstractions. Abstraction is not the same at each phase of a design, and resolution of implementation details is deferred in the design process until such details are really needed. The use of such abstractions reduces work because existing representations can be reused, and different representations can be related through their common pieces. Key to developing these abstractions is the exploitation of hierarchical composability (i.e., the construction of complex standards or information models from simpler ones). Researchers should seek appropriate abstractions for mechanical products and develop tools to support design in terms of those abstractions.

In electronic design, many sophisticated tools have been developed to support physical aspects of design, while fewer tools exist for aspects of conceptual design such as requirement gathering or architectural decision making. This disparity suggests that it is far easier to collect, characterize, and represent data about "things" than about ideas and decisions. Nevertheless, the payoff for automated support of activities such as requirement gathering, corporate decision making, and overall product architecture is so high that research directed at these targets is also worth undertaking.

The committee believes that concentrating on the areas described in the rest of this chapter will yield the most significant benefits to the IPPD process in the short to medium term. Other areas in the IPPD process not described in this chapter (mostly in the conceptual design area) will be advanced in the short to medium term by better communication capabilities and by better and additional access to data.

The committee notes also that success in improving design is likely to yield particular benefits to small manufacturers. To a considerable degree, tools that support product and process design can be categorized as "mostly software"; that is, they require low capital investment to obtain. Moreover, the increasing power of computational hardware (and its dropping cost) will make these tools more accessible to manufacturing firms with small capital budgets. To capitalize on this enabling trend, it is necessary to design these tools and their interfaces so that small businesses with limited technical depth can use them readily. A growing information infrastructure (Chapter 6) will link these small firms with each other and with larger firms. The emergence of standard data formats and interoperable tools will facilitate the spread of advanced capabilities.

Research for Product Description

Communication among engineering professionals has always relied on models—sketches, drawings, analytic models of behavior, or many other symbolic representations of knowledge related to products. However, what distinguishes data or information models from arbitrary documentation, such as simple reports or drawings, is the addition of formalization. This formalization is motivated by the need for unambiguous communication between collaborators and/or our desire to use and interoperate among a growing set of computer-based application tools.

Specifically, a data model is expressed in some data description language. A language specification defines the form and meaning (syntax and semantics) of entities in the language. The initial graphics exchange specification (IGES; IGES, 1993), for instance, specifies the language of IGES data files by specifying their syntactic form and relating the entities in the data files to known geometric entities.

Traditionally, the specifications of data description languages such as IGES have been written in a natural language such as English, and a human being must use that description to build a software parser/recognizer for the data description language. The semantics of the elements of the language remain expressed solely in English. More recent efforts such as PDES/STEP (product data exchange using the standard for the exchange of product model data; ISO, 1994) provide a more formal language, EXPRESS, for use as a data description language in which data models are described. EXPRESS allows the modeler to capture some of the semantics of the data by explicitly recognizing relationships between data elements along with the cardinality of such relationships and by capturing constraints between data elements.

Box 3.3 describes two approaches to knowledge representation. Appropriate formalisms also support the generation of agreement on queries about data; application tool development; language translation; and services such as change notification, management of information dependencies, and matching of information producers and consumers.

A highly general expressive capability is needed to support the exchange of engineering knowledge, specifically the ability to exchange a conceptualization that specifies the objects that are presumed to exist in some area of interest, as well as the relationships that hold among them (Genesereth and Nilsson, 1987). For example, for two parties to discuss models of dynamical behavior, they must first agree on the use of terms such as generalized coordinates, position, and velocity, and they must understand the laws on which the models rely (e.g., Newton's laws, Lagrange's equations, or Kane's equations). At the same time, increasingly on information models to help bridge the gap between multidisciplinary users of diverse tools; exchangeable geometric models are thus a particularly pressing need. Success in developing exchangeable representations of performance, geometry, and process requirements is a prerequisite for their use in design tools and by practitioners in the allied domains of process equipment design and shop floor planning and operations, as well as by designers in other companies or in other technical domains.

Research for Process Description

Tools for process description have application to the design and operation of factories; in addition, they are the basis for experiments with and evaluation of control and organizational changes before actual systems are installed. Process descriptions will also be used to enhance product design, so that by simulation the best process can be matched to the product design (and vice versa) for maximum economic advantage (or to satisfy whatever criteria—such as quality or time to delivery—are important for the particular case). Box 3.7 describes a rich and productive interaction among process design, product design, and what happens on the shop floor.

The primary need in process description is formalization, which is necessary for representing processes in sufficient detail and with enough specificity to make the process description adequately complete and unambiguous. Such formalisms allow designers to describe, enforce, and simulate processes, including factory processes (involving both machines and people), design activities, and decision processes.

Research needs in the area of process description include:

- A language for expressing process descriptions that facilitates checking for correctness and completeness and the capability to express not only nominal process behavior but also variant behavior. Such a language must also be translatable across technical domains.

- Process model representation schemes (both aggregate and detailed) and on-line data collection. Data included in such schemes should describe logistic, fabrication, material-handling, inspection, assembly, and test processes and should give information on the characteristics, capabilities, and costs of various production and assembly methods. Data should be captured as a product is being produced so that the process can be improved.
- Specific process models that reflect all relevant spatial and temporal transformations. Such models are critical for the local control and planning of manufacturing operations and will draw on knowledge about the kinematic capabilities of individual pieces of equipment and other process limitations, processing capabilities of the equipment, and tool and fixturing capabilities associated with the equipment. Ultimately, these models should contain the detail necessary for dynamic control of the individual operations as well as the information required to simulate the operation of the manufacturing system, indicate the effects of perturbing the operational parameters as well as the effects of complex interactions among processes, and be generalizable across a wide range of production environments by appropriate parameterization.
- Algorithms and tools to solve process problems. For example, one such problem is the determination of efficient assembly sequences; an inappropriate assembly sequence may result in the need for a tool to reorient the item being assembled many more times than necessary, thus increasing the time needed for assembly and the likelihood of breakage. Such factors are to a considerable degree irrelevant to the design of the product itself but have a significant influence on the cost of making the product. Other important process problems include the determination of efficient equipment layouts on a factory floor, equipment selection (matching equipment capability to process needs), make-buy decisions, and determination of how best to cut and shape materials to minimize waste.
- Dynamic models for describing resources available over time to the manufacturing system. Such models will be used in both the design and the operational control of manufacturing systems. Common representations and descriptions of resources are necessary to enable development of transferable (from planning to analysis to control) models and analysis. Despite the importance of resource management, little research and effort have been devoted to creating generic representations of resources. As a result, specific resource characteristics must be recreated each time a modeling activity is undertaken. Resources include those related to fabrication (e.g., tooling, machines, available controller features) and their

interconnection, as well as system resources such as corporate information or knowledge and information such as company or external standards.

RESEARCH FOR TOOLS TO SUPPORT INTEGRATED PRODUCT AND PROCESS DESIGN

Computer tools that directly aid the management of the IPPD process itself would be helpful to managers. Today, such tools are limited primarily to communication aids or "groupware" for helping people post notices and share information. Means of describing and managing the design process need to be developed. Few tools exist for creating, monitoring, and guiding the design process itself, except for familiar project management tools like PERT. (PERT is largely a schedule and resource management tool.) Scheduling aids beyond PERT are required to help in determining effective task sequences, setting up information flows, establishing schedules and milestones, identifying people, assigning work to them, routing information to them, and linking them to colleagues elsewhere. Existing tools do not help to identify information flows or facilitate them.

A single design decision may have several simultaneous impacts, some of which may be beneficial and others adverse. The design environment should support techniques to express comparisons and trade-offs vividly so that a designer can assess the impact of a wide range of design decisions on a product's cost, time to produce, or quality. Tools are needed that focus on the identification of design trade-offs between cost, performance, and reliability; alternate space allocations; functional decomposition; subassembly definition; three-dimensional geometric reasoning; and make-or-buy decisions.

To produce such tools, research is needed on decision tools that draw on the product-process data model and performance simulations of the product being designed, as well as process models and various data on costs and tolerances of different processes. These decision tools will sort data and models on the basis of criteria supplied by the designer to aid in making comparisons between alternate designs and processes.

In addition, improvements to simulation and rapid prototyping tools should be brought to bear on the problems of viewing physical parts, "visualizing" complex relationships (certainly between physical parts, but also between more abstract relationships such as design requirements and their costs), and presenting design alternatives to customers and to designers. This category also includes methods or tools to handle groups of parts such as assemblies, subsystems, product families, made-to-order configurations, and selected combinations of parts that create different product models by virtue of which parts are selected. (Box 3.7 describes such an application.)

Research is needed on design methods and computer tools adapted for the design and creation of groups of parts or systems, in addition to individual parts. Such methods and tools will enable the designer to divide a product into subassemblies, design optimal in-process test strategies during assembly, and identify assembly sequences that minimize cost, tolerance errors, material handling, and part damage during assembly. Also, the methods and tools will include standard design modules and methods that facilitate optimization of part or all of a product for cost and quality. Such optimization requires a deep understanding of specific components and features of the product.

Finally, it would be highly desirable to have tools that would allow product design and process design to proceed more in parallel. These tools would enable product designers to work with some degree of incomplete information about the process designer's work, and vice versa. Successful development of such tools would contribute greatly to the reduction of needed design time.

RESEARCH AREAS NOT SPECIFIC TO MANUFACTURING

Geometric Reasoning

A generic intellectual activity required by mechanical design is geometric reasoning. A major difference between VLSI design and mechanical design is the degree to which three-dimensional geometric reasoning is fundamental to even the simplest mechanical designs; VLSI design is based on two-dimensional (2-D) analysis (or at worst, 2 1/2 dimensional analysis—the use of stacked 2-D layers). Three factors make three-dimensional (3-D) reasoning more complex: 3-D design is more likely to involve moving parts or flows; 3-D design may involve interconnections and tolerance relationships between 2-D domains; and 3-D designs can be visualized only in cross sections, perspective, and exploded views. Research is needed on building robust geometric modelers. Many of the "boundary representation" modelers in use today are not robust: interactions between the algorithms they use and finite-precision arithmetic offered by the computer result in certain modeling operations that yield incorrect results. Truly robust algorithms to correct such problems remain a challenging research problem. Ultimately, the design environment should support improved visualization tools or other design aids that will help make geometric reasoning faster and efficiently achievable by a broader range of people. In addition, tools that undertake geometric reasoning automatically (i.e., without relying on a skilled human) may be able to replace human designers for certain purposes.

Knowledge and Information Management

Basic to design are many issues of data management and of data themselves. The future design environment will include a number of data management methods and tools. For example, the design environment will have to handle data legacy issues, such as converting data from one CAD system to another and preserving old data for decades or more so that they can still be read, edited, and processed. Today, such data are either lost, kept on paper, or accessed in a limited way by old hardware kept on hand for the purpose.

A second data management issue is that the design environment will have to include ways to capture corporate memory and knowledge so that successors of current designers can tell what knowledge was used, what competitive methods were used, what errors were made, and on what factors success was based. The ability to record design history and rationale is of particular importance. Every design is a historical web of decisions that grow out of each other and depend on each other. Revision, whether for correcting an error, absorbing a new outside circumstance, or improving manufacturability, requires unraveling the web to a certain degree. The design selected, the web of decisions leading to it, and "roads not taken" indicate the corporate state of belief at the time the decisions were made and thus form a historic context.

Specific Research Questions

A variety of specific research questions are motivated by discussions earlier in this chapter. In accordance with the charge of the study, the focus of these research questions is largely information technology and associated disciplines such as mathematics. If the underlying science and engineering aspects of manufacturing are well understood, IT can help immeasurably in exploiting such knowledge and information, but by itself IT is not a substitute for that knowledge.

- How should the data contained in a product data model be organized to accommodate their huge size and complexity and the many disciplines that need access to them?
- How much of that information is physical and how much is "relational"? How much can be captured in traditional geometry and how much is nongeometric or not focused on one item but shared or spread among many items, or even not attached to specific items?
- To what extent can a product description for mechanical items be converted automaticallyinto a production plan, that is, a sequence of fabrication steps that transform raw materials into a final product?
- Can a general process description language be developed that would be both man- and machine-intelligible, permitting processes to be described more precisely than is possible now?
- How can CAD tools be used at different stages of the design process? Can high-order abstractions be used as a starting point for encapsulating product and process facts and knowledge, geometry, requirements, tolerances, and other product characteristics, providing a link between product function, geometry, and processes? Can

"meta-features" be defined that will encapsulate groups of features, creating a feature hierarchy? If not, how can the scale of real designs be encompassed using features or any other means of capturing and combining detailed design data, geometry, and intent?

- Features are often very process-dependent, and the fabrication of parts may require the application of multiple processes, each of which uses a different set of features to address the same part. Attaching features to descriptions of parts may prove impractical due to the differing feature sets for the same part. Feature-based product description may prove unproductive. In this event, to what extent is it feasible to design process models that interact with separate product models to generate appropriate product models?

How can the human interfaces of CAD systems be improved so that more complex shapes, assemblies, and other multidimensional problems can be handled more easily?

- What new process analysis tools can be developed, especially to handle complex problems like assembly, model mix manufacture, and tolerances?
- How can the reliability of new product or process designs be better predicted, including trade-offs between cost and reliability?
- What new languages or data structures can be developed to better describe product requirements, such as performance, reliability, shapes, interconnections, interfaces, and tolerances?
- Similarly, what new languages or data structures can be developed to better describe process requirements, such as performance, cost, reliability, ease of diagnosis and repair, material handling, ease of use, and ease of modification?
- What new data logging and correlation methods can be devised that would help the process of continuous improvement, such as finding multiple occurrences of the same type of machine failure or deducing what is the best sequence for testing a broken system to diagnose its problems?

CREATING ECO-EFFICIENT PRODUCTS AND PROCESSES

The central concept developed in the book is concept of Design for Environment (DfE), which originated in 1992 with the efforts of many electronics firms that attempted to build environmental awareness into their product development tasks. DfE is defined as the "systematic consideration of design performance with respect to environmental, health, and safety objectives over the full product and process life cycle." The driving forces behind DfE include the fact that costumers have become increasingly concerned about the environmental friendliness of the products they acquire. In addition, the International Organization for Standardization (ISO) has developed ISO 14000 standards for environmental management systems and many government agencies took aggressive steps to assure that manufacturers are responsible for recovery of products and materials at the end of their useful lives. Similar efforts have been put forward by consortia as the Global Environmental Management Initiative and the Business Council for Sustainability Development.

As Finksel points out "DfE is at the crossroads of two trusts that are transforming the nature of manufacturing in the world", and this is what makes this concept interesting from a systems thinking perspective. DfE combines and results from the interaction between two systems, the economics system and the environmental system and branches out into two areas of action: sustainable development, which aims at designing industrial progress that "meets the needs of the present without compromising the ability of future generations to meet their own needs"; and enterprise integration, which is concerned with factoring environmental concerns into the profitability and effectiveness of the firm. Sustainable development is a statement more related to a macro view of the industrial world, whereas enterprise integration is more concerned about how to develop and integrate sustainability requirements at the micro level.

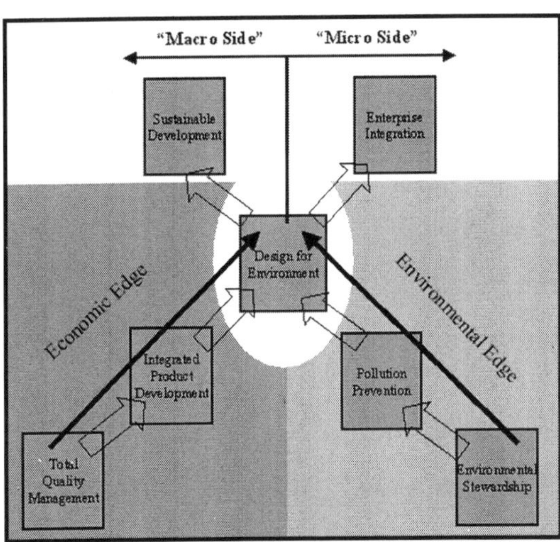

Fig. Conceptualizing Design for Environment (DfE) (Adapted from Fiksel, 96). DfE results from the combination of economic and environmental performance and branches out to areas of research: sustainable development that takes a macro view of the problem and enterprise integration that deals with implementing DfE at the micro level.

Once the premise of DfE is understood, the big challenge relies on its implementation. Barriers to this process are the lack of expertise among product development engineers to take into account environmental concerns, the complexity of environmental phenomena, which makes it difficult to build "systems dynamics" type of models to understand them and the fact that the economic environment in which products are produced and used is much more complex and difficult to control than the products themselves.

Finksel's book sheds light into these issues. The second part of the book is more associated with the first difficulty referred above and is instructive about how one can look and adapt product development to environmental concerns. The third part of the book addresses the other two challenges presenting a series of cases studies that serve as a reference to successfully implement DfE. There is also an introductory part, which makes the case for environmental concerns, and a fourth part that assembles a set of suggestions for the future. The structure of the book is depicted in Figure along with a timeline that one can associate to its contents.

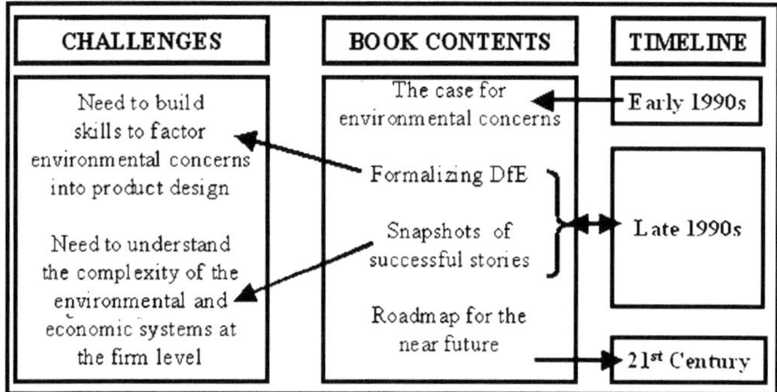

Fig. The main challenges to implementing DfE and an index of the contents of the book that aim at addressing those concerns. A timeline associated with the structure of the book relates it to the broader context of research to accomplish eco-efficiency.

The second part of the book includes chapters that formalize the principles of DfE, chapters on how to integrate DfE into more comprehensive life-cycle management strategies and chapters about metrics for assessing environmental performance and improvement. From a systems thinking perspective these chapters deal mostly with the question of where one should draw the boundary of the system when talking about product development. This is clear from the effort of building up metrics for performance that take into account environmental variables and factoring these into the manufacturing processes. In other words, the boundary of the system is placed allowing for retrofitting the externalities of both the manufacturing processes and the manufactured products on the environment into the design of the product development process.

Central to ILCM is the idea of Life-Cycle Cost Management (LCCM) and definition the life-cycle cost as the sum of the cost of acquisition and the cost of use with the cost of disposal and post-disposal. This is the mechanism that society, and particularly the government, uses to take care of the problem: penalize someone for environment degradation and that someone will have a "stimulus "to do something about it. Furthermore, making firms responsible for their own waste does not seem a bad idea in the first place because it punishes firms proportionally to the negative effects they cause.

Left with this equation, the reader can think of environmental concerns as a source of increased costs. But Fiksel unravels the mystery of the win-win situations with a series of real world examples, which he calls "Pioneer Stories". The books presents 11 stories of successful applications of DfE, in chapters written by invited people from companies spanning from the computer industry, to telecommunications, semiconductor, gas and electricity, healthcare, beverages and chemicals. This broad choice of industries is not by accident. Clearly, Fiksel also tries to make the case for eco-efficiency by showing that win-win scenarios are not specific to a particular manufacturing process of product but rather apply to every corner of the economy.

These chapters are very similar in spirit. The introduction of DfE into Microelectronics and Computer Technology Corporation (MCC), based in Austin, TX and it describes how one can take spot win-win situations for the case of manufacturing a workstation. This firm conducted a study about the manufacturing process of a workstation. They focused on four areas: energy, materials, waste and water. They have found that Printed Wiring Boards (PWBs) consume a significant amount of energy and water along the manufacturing process. This study has stimulated a number of activities within MCC, within the government and in the industry at large. A roadmap providing directions for greening the electronics industry was published two years later and EPA's Environmental Technology Initiative developed a DfE project to further study PWBs. This project compared various fabrication approaches and identified alternative technologies.

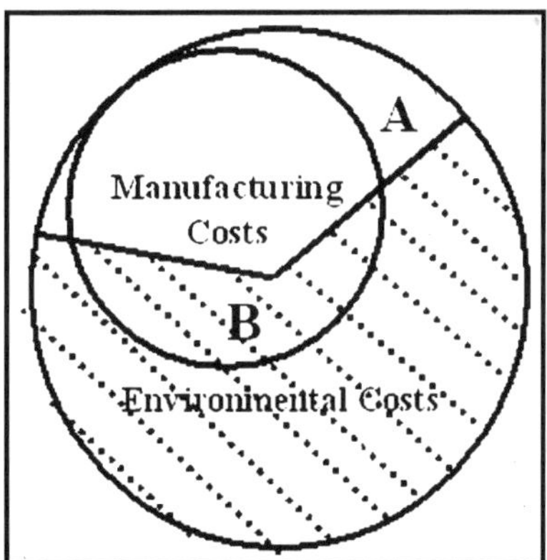

Fig. Illustration of how net benefits may be accomplished from considering environmental concerns and technological innovation. The inner circle represents the original manufacturing costs. When environmental related costs are added the total costs amount to the outer circle. But environmental concerns may trigger technological innovation that cuts costs, as represented by the dashed area. Net benefits are achieved when A is smaller than B.

This is a common fact among most of the cases reported in this book: DfE often triggers technological innovation as a way to lessen negative externalities to the environment. Once firms factor costs from affecting the environment into their decision processes, savings associated with those costs usually come tied to other types of cost savings. In other words, reducing the costs associated with the environment often induces technological innovations that also cut costs in other tasks of the manufacturing process. Net benefits emerge when the benefits of reducing the overall cost structure overweight the losses of incorporating environmental related costs.

NON TRADITIONAL MANUFACTURING PROCESSES

Non-traditional manufacturing processes is defined as a group of processes that remove excess material by various techniques involving mechanical, thermal, electrical or chemical energy or combinations of these energies but do not use a sharp cutting tools as it needs to be used for traditional manufacturing processes. Extremely hard and brittle materials are difficult to machine by traditional machining processes such as turning, drilling, shaping and milling. Non traditional machining processes, also called advanced manufacturing processes, are employed where traditional machining processes are not feasible, satisfactory or economical due to special reasons as outlined below.

- Very hard fragile materials difficult to clamp for traditional machining
- When the workpiece is too flexible or slender
- When the shape of the part is too complex.

Several types of non-traditional machining processes have been developed to meet extra required machining conditions. When these processes are employed properly, they offer many advantages over non-traditional machining processes. The common non-traditional machining processes are described in this section.

ELECTRICAL DISCHARGE MACHINING (EDM)

Fig. Electrical discharge machine

Electrical discharge machining (EDM) is one of the most widely used non-traditional machining processes. The main attraction of EDM over traditional machining processes such as metal cutting using different tools and grinding is that this technique utilises thermoelectric process to erode undesired materials from the workpiece by a series of discrete electrical sparks between the workpiece and the electrode. A picture of EDM machine in operation is shown in Figure.

The traditional machining processes rely on harder tool or abrasive material to remove the softer material whereas non-traditional machining processes such as EDM uses electrical spark or thermal energy to erode unwanted material in order to create desired shape. So, the hardness of the material is no longer a dominating factor for EDM process. A schematic of an EDM process, where the tool and the workpiece are immersed in a dielectric fluid.

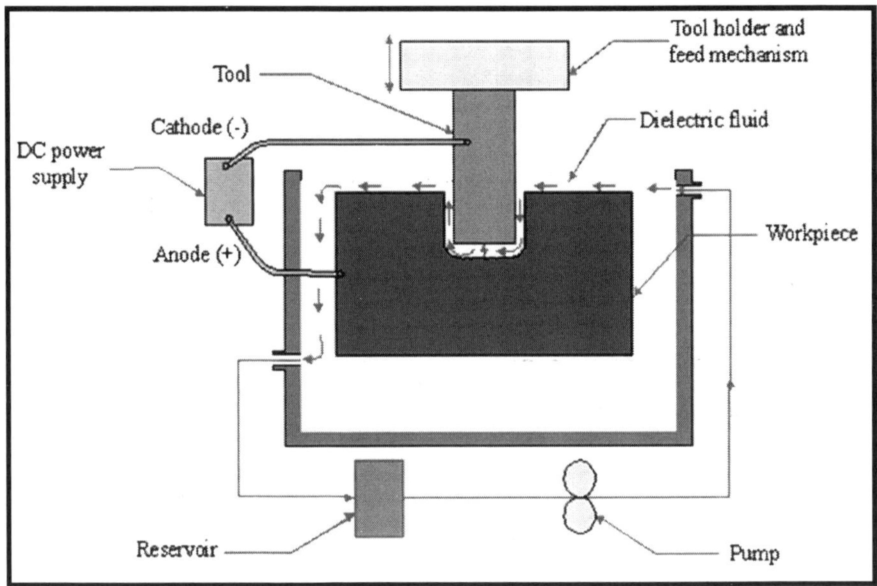

Fig. Schematic of EDM process

EDM removes material by discharging an electrical current, normally stored in a capacitor bank, across a small gap between the tool (cathode) and the workpiece (anode) typically in the order of 50 volts/10amps.

APPLICATION OF EDM

The EDM process has the ability to machine hard, difficult-to-machine materials. Parts with complex, precise and irregular shapes for forging, press tools, extrusion dies, difficult internal shapes for aerospace and medical applications can be made by EDM process. Some of the shapes made by EDM process.

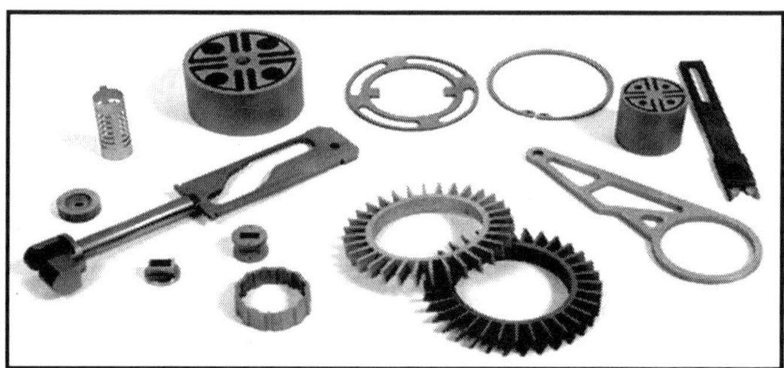

Fig. Difficult internal parts made by EDM process

Working principle of EDM

The beginning of EDM operation, a high voltage is applied across the narrow gap between the electrode and the workpiece. This high voltage induces an electric field in the insulating dielectric that is present in narrow gap between electrode and workpiece. This cause conducting particles suspended in the dielectric to concentrate at the points of strongest electrical field. When the potential difference between the electrode and the workpiece is sufficiently high, the dielectric breaks down and a transient spark discharges through the dielectric fluid, removing small amount of material from the workpiece surface. The volume of the material removed per spark discharge is typically in the range of 10-6 to 10-6 mm3.

The material removal rate, MRR, in EDM is calculated by the following foumula:

MRR = 40 I/ Tm 1.23 (cm3/min)

Where, I is the current amp,

Tm is the melting temperature of workpiece in 0C'

Advantages of EDM

The main advantages of DM are:

- By this process, materials of any hardness can be machined;
- No burrs are left in machined surface;
- One of the main advantages of this process is that thin and fragile/ brittle components can be machined without distortion;
- Complex internal shapes can be machined.

Limitations of EDM

The main limitations of this process are:

- This process can only be employed in electrically conductive materials;
- Material removal rate is low and the process overall is slow compared to conventional machining processes;
- Unwanted erosion and over cutting of material can occur;
- Rough surface finish when at high rates of material removal.

Dielectric fluids

Dielectric fluids used in EDM process are hydrocarbon oils, kerosene and deionised water. The functions of the dielectric fluid are to:

- Act as an insulator between the tool and the workpiece.
- Act as coolant.
- Act as a flushing medium for the removal of the chips.

The electrodes for EDM process usually are made of graphite, brass, copper and copper-tungsten alloys.

Design considerations for EDM process are as follows:

- Deep slots and narrow openings should be avoided.
- The surface smoothness value should not be specified too fine.
- Rough cut should be done by other machining process. Only finishing operation should be done in this process as MRR for this process is low.

Wire EDM

EDM, primarily, exists commercially in the form of die-sinking machines and wire-cutting machines (Wire EDM). In this process, a slowly moving wire travels along a prescribed path and removes material from the workpiece. Wire EDM uses electro-thermal mechanisms to cut electrically conductive materials. The material is removed by a series of discrete discharges between the wire electrode and the workpiece in the presence of dieelectirc fluid, which creates a path for each discharge as the fluid becomes ionized in the gap. The area where discharge takes place is heated to extremely high temperature, so that the surface is melted and removed. The removed particles are flushed away by the flowing dielectric fluids.

The wire EDM process can cut intricate components for the electric and aerospace industries. This non-traditional machining process is widely used to pattern tool steel for die manufacturing.

Fig. Wire erosion of an extrusion die

The wires for wire EDM is made of brass, copper, tungsten, molybdenum. Zinc or brass coated wires are also used extensively in this process. The wire used in this process should posses high tensile strength and good electrical conductivity. Wire EDM can also employ to cut cylindrical objects with high precision.

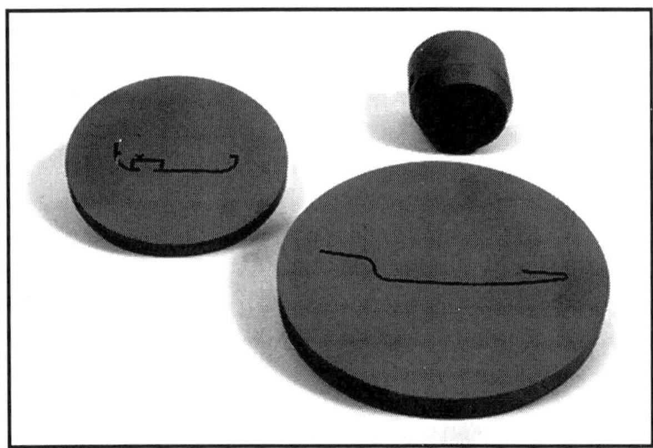

Fig. Sparked eroded extrusion dies

This process is usually used in conjunction with CNC and will only work when a part is to be cut completely through. The melting temperature of the parts to be machined is an important parameter for this process rather than strength or hardness.

The surface quality and MRR of the machined surface by wire EDM will depend on different machining parameters such as applied peak current, and wire materials.

CHEMICAL MACHINING (CM)

Chemical machining (CM) is the controlled dissolution of workpiece material (etching) by means of a strong chemical reagent (etchant). In CM material is removed from selected areas of workpiece by immersing it in a chemical reagents or etchants; such as acids and alkaline solutions. Material is removed by microscopic electrochemical cell action, as occurs in corrosion or chemical dissolution of a metal.

This controlled chemical dissolution will simultaneously etch all exposed surfaces even though the penetration rates of the material removal may be only 0.0025–0.1 mm/min.

The basic process takes many forms: chemical milling of pockets, contours, overall metal removal, chemical blanking for etching through thin sheets; photochemical machining (pcm) for etching by using of photosensitive resists in microelectronics; chemical or electrochemical polishing where weak chemical reagents are used (sometimes with remote electric assist) for polishing or deburring and chemical jet machining where a single chemically active jet is used.

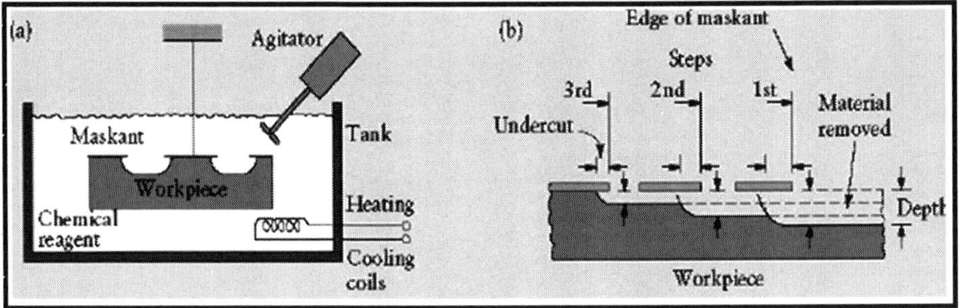

Fig. (a) Schematic of chemical machining process (b) Stages in producing a profiled cavity by chemical machining (Kalpakjain & Schmid)

CHEMICAL MILLING

In chemical milling, shallow cavities are produced on plates, sheets, forgings and extrusions. The two key materials used in chemical milling process are etchant and maskant. Etchants are acid or alkaline solutions maintained within controlled ranges of chemical composition and temperature. Maskants are specially designed elastomeric products that are hand strippable and chemically resistant to the harsh etchants.

Steps in chemical milling

- Residual stress relieving: If the part to be machined has residual stresses from the previous processing, these stresses first should be relieved in order to prevent warping after chemical milling.
- Preparing: The surfaces are degreased and cleaned thoroughly to ensure both good adhesion of the masking material and the uniform material removal.
- Masking: Masking material is applied (coating or protecting areas not to be etched).
- Etching: The exposed surfaces are machined chemically with etchants.
- Demasking: After machining, the parts should be washed thoroughly to prevent further reactions with or exposure to any etchant residues. Then the rest of the masking material is removed and the part is cleaned and inspected.

ELECTROCHEMICAL MACHINING (ECM)

Electrochemical machining (ECM) is a metal-removal process based on the principle of reverse electroplating. In this process, particles travel from the anodic material (workpiece) toward the cathodic material (machining tool). A current of electrolyte fluid carries away the deplated material before it has a chance to reach the machining tool. The cavity produced is the female mating image of the tool shape.

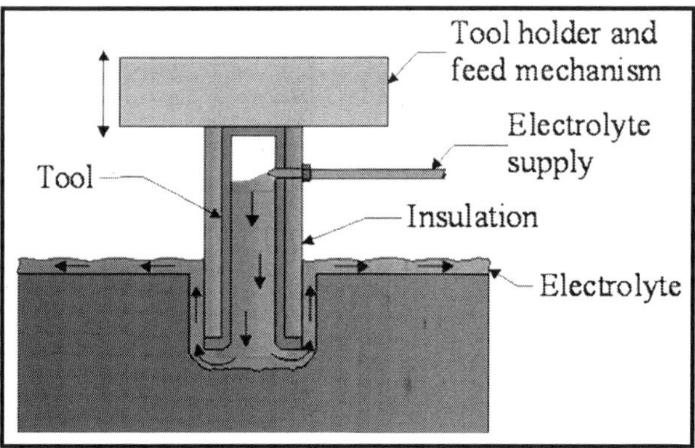

Fig. ECM process

Similar to EDM, the workpiece hardness is not a factor, making ECM suitable for machining difficult-to –machine materials. Difficult shapes can be made by this process on materials regardless of their hardness. A schematic representation of ECM process. The ECM tool is positioned very close to the workpiece and a low voltage, high amperage DC current is passed between the workpiece and electrode. Some of the shapes made by ECM process.

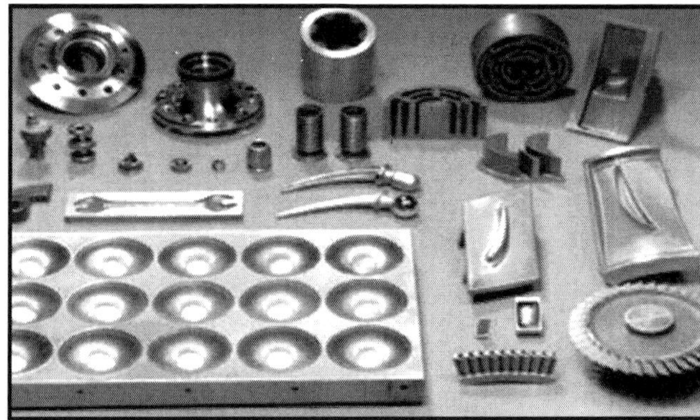

Fig. Parts made by ECM

ADVANTAGES OF ECM

- The components are not subject to either thermal or mechanical stress.
- No tool wear during ECM process.
- Fragile parts can be machined easily as there is no stress involved.
- ECM deburring can debur difficult to access areas of parts.
- High surface finish (up to 25 μm in) can be achieved by ECM process.
- Complex geometrical shapes in high-strength materials particularly in the aerospace industry for the mass production of turbine blades,

jet-engine parts and nozzles can be machined repeatedly and accurately.
- Deep holes can be made by this process.

Limitations of ECM

- ECM is not suitable to produce sharp square corners or flat bottoms because of the tendency for the electrolyte to erode away sharp profiles.
- ECM can be applied to most metals but, due to the high equipment costs, is usually used primarily for highly specialised applications.

Material removal rate, MRR, in electrochemical machining:

- MRR = C.I. h (cm3/min)
- *C:* Specific (material) removal rate (e.g., 0.2052 cm3/amp-min for nickel);
- *I:* Current (amp);
- *h:* Current efficiency (90–100%).

The rates at which metal can electrochemically remove are in proportion to the current passed through the electrolyte and the elapsed time for that operation. Many factors other than current influence the rate of machining. These involve electrolyte type, rate of electrolyte flow, and some other process conditions.

ULTRASONIC MACHINING (USM)

USM is mechanical material removal process or an abrasive process used to erode holes or cavities on hard or brittle workpiece by using shaped tools, high frequency mechanical motion and an abrasive slurry. USM offers a solution to the expanding need for machining brittle materials such as single crystals, glasses and polycrystalline ceramics, and increasing complex operations to provide intricate shapes and workpiece profiles. It is therefore used extensively in machining hard and brittle materials that are difficult to machine by traditional manufacturing processes. The hard particles in slurry are accelerated toward the surface of the workpiece by a tool oscillating at a frequency up to 100 KHz - through repeated abrasions, the tool machines a cavity of a cross section identical to its own. A schematic representation of USM.

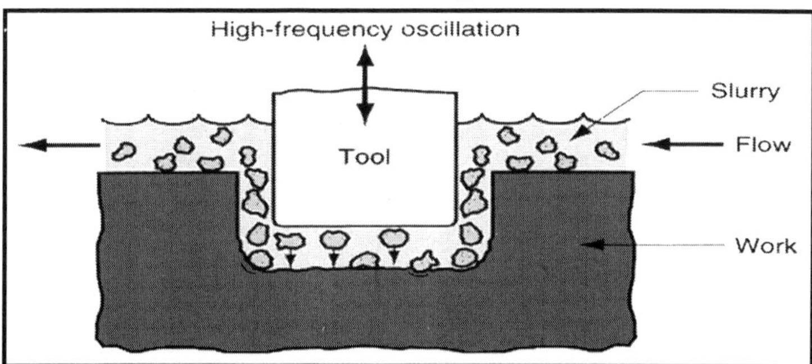

Fig. Schematic of ultrasonic machine tool

USM is primarily targeted for the machining of hard and brittle materials (dielectric or conductive) such as boron carbide, ceramics, titanium carbides, rubies, quartz etc. USM is a versatile machining process as far as properties of materials are concerned. This process is able to effectively machine all materials whether they are electrically conductive or insulator.

For an effective cutting operation, the following parameters need to be carefully considered:

- The machining tool must be selected to be highly wear resistant, such as high-carbon steels.
- The abrasives (25-60 mm in dia.) in the (water-based, up to 40% solid volume) slurry includes: Boron carbide, silicon carbide and aluminum oxide.

APPLICATIONS

The beauty of USM is that it can make non round shapes in hard and brittle materials. Ultrasonically machined non round-hole part.

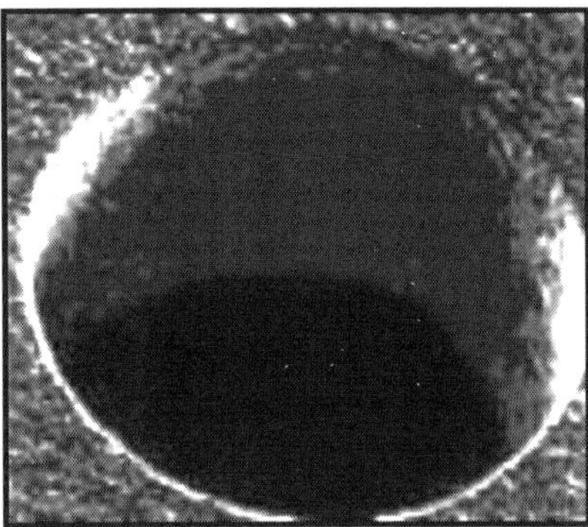

Fig. A non-round hole made by USM

ADVANTAGE OF USM

USM process is a non-thermal, non-chemical, creates no changes in the microstructures, chemical or physical properties of the workpiece and offers virtually stress free machined surfaces.

- Any materials can be machined regardless of their electrical conductivity
- Especially suitable for machining of brittle materials
- Machined parts by USM possess better surface finish and higher structural integrity.
- USM does not produce thermal, electrical and chemical abnormal surface

Some disadvantages of USM

- USM has higher power consumption and lower material-removal rates than traditional fabrication processes.

- Tool wears fast in USM.
- Machining area and depth is restraint in USM.

LASER–BEAM MACHINING (LBM)

Laser-beam machining is a thermal material-removal process that utilizes a high-energy, coherent light beam to melt and vaporize particles on the surface of metallic and non-metallic workpieces. Lasers can be used to cut, drill, weld and mark. LBM is particularly suitable for making accurately placed holes. A schematic of laser beam machining.

Different types of lasers are available for manufacturing operations which are as follows:

- CO_2 (pulsed or continuous wave): It is a gas laser that emits light in the infrared region. It can provide up to 25 kW in continuous-wave mode.
- Nd:YAG: Neodymium-doped Yttrium-Aluminum-Garnet (Y3A15O12) laser is a solid-state laser which can deliver light through a fibre-optic cable. It can provide up to 50 kW power in pulsed mode and 1 kW in continuous-wave mode.

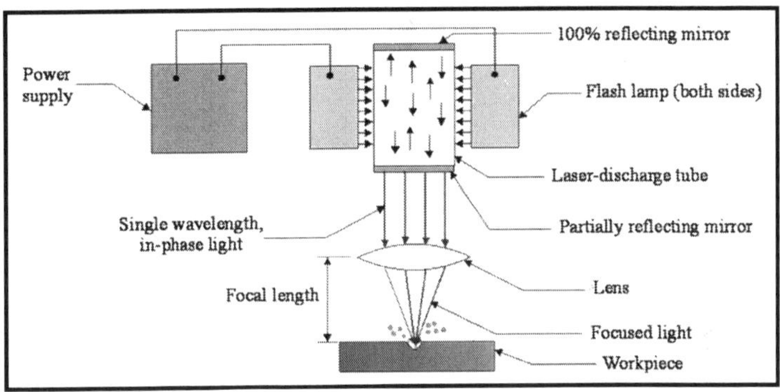

Fig. Laser beam machining schematic

APPLICATIONS

LBM can make very accurate holes as small as 0.005 mm in refractory metals ceramics, and composite material without warping the workpieces. This process is used widely for drilling and cutting of metallic and non-metallic materials. Laser beam machining is being used extensively in the electronic and automotive industries.

Laser beam cutting (drilling)

- In drilling, energy transferred (e.g., via a Nd:YAG laser) into the workpiece melts the material at the point of contact, which subsequently changes into a plasma and leaves the region.
- A gas jet (typically, oxygen) can further facilitate this phase transformation and departure of material removed.

- Laser drilling should be targeted for hard materials and hole geometries that are difficult to achieve with other methods.

A typical SEM micrograph hole drilled by laser beam machining process employed in making a hole.

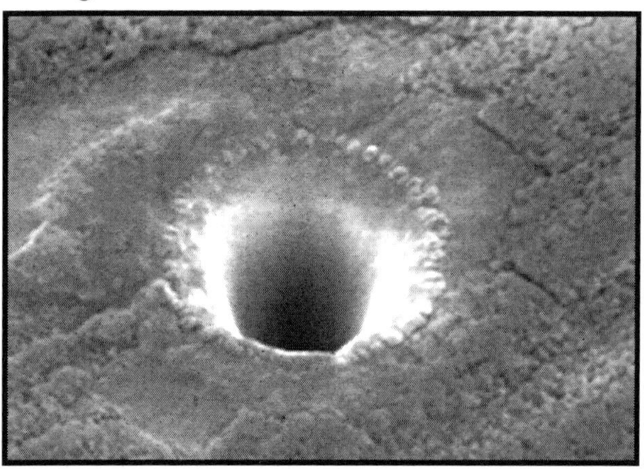

Fig. SEM micrograph hole drilled in 250 micro meter thick Silicon Nitride with 3rd harmonic Nd: YAG laser

Laser beam cutting (milling)

- A laser spot reflected onto the surface of a workpiece travels along a prescribed trajectory and cuts into the material.
- Continuous-wave mode (CO2) gas lasers are very suitable for laser cutting providing high-average power, yielding high material-removal rates, and smooth cutting surfaces.

Advantage of laser cutting

- No limit to cutting path as the laser point can move any path.
- The process is stress less allowing very fragile materials to be laser cut without any support.
- Very hard and abrasive material can be cut.
- Sticky materials are also can be cut by this process.
- It is a cost effective and flexible process.
- High accuracy parts can be machined.
- No cutting lubricants required
- No tool wear
- Narrow heat effected zone

Limitations of laser cutting

- Uneconomic on high volumes compared to stamping
- Limitations on thickness due to taper
- High capital cost

- High maintenance cost
- Assist or cover gas required

WATER JET CUTTING

Water jet cutting can reduce the costs and speed up the processes by eliminating or reducing expensive secondary machining process. Since no heat is applied on the materials, cut edges are clean with minimal burr. Problems such as cracked edge defects, crystalisation, hardening, reduced wealdability and machinability are reduced in this process.

Water jet technology uses the principle of pressurising water to extremely high pressures, and allowing the water to escape through a very small opening called "orifice" or "jewel". Water jet cutting uses the beam of water exiting the orifice to cut soft materials. This method is not suitable for cutting hard materials. The inlet water is typically pressurised between 1300 – 4000 bars. This high pressure is forced through a tiny hole in the jewel, which is typically o.18 to 0.4 mm in diameter. A picture of water jet machining process.

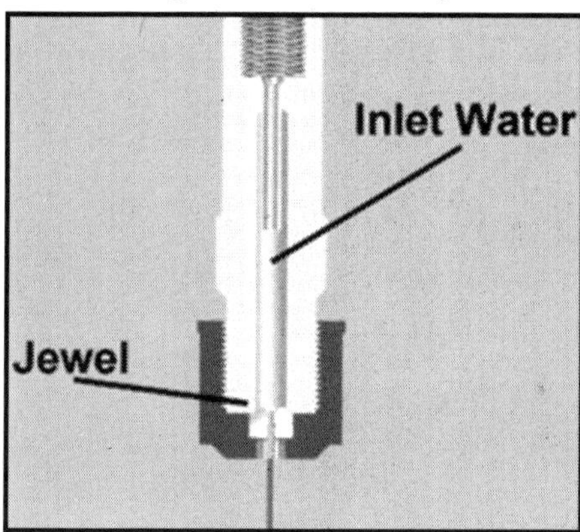

Fig. Water jet cutting

Applications

Water jet cutting is mostly used to cut lower strength materials such as wood, plastics and aluminium. When abrasives are added, (abrasive water jet cutting) stronger materials such as steel and tool steel can be cut.

Advantages of water jet cutting

- There is no heat generated in water jet cutting; which is especially useful for cutting tool steel and other metals where excessive heat may change the properties of the material.
- Unlike machining or grinding, water jet cutting does not produce any dust or particles that are harmful if inhaled.
- Other advantages are similar to abrasive water jet cutting

Disadvantages of water jet cutting

- One of the main disadvantages of water jet cutting is that a limited number of materials can be cut economically.
- Thick parts cannot be cut by this process economically and accurately
- Taper is also a problem with water jet cutting in very thick materials. Taper is when the jet exits the part at different angle than it enters the part, and cause dimensional inaccuracy.

ABRASIVE WATER-JET CUTTING

Abrasive water jet cutting is an extended version of water jet cutting; in which the water jet contains abrasive particles such as silicon carbide or aluminium oxide in order to increase the material removal rate above that of water jet machining. Almost any type of material ranging from hard brittle materials such as ceramics, metals and glass to extremely soft materials such as foam and rubbers can be cut by abrasive water jet cutting. The narrow cutting stream and computer controlled movement enables this process to produce parts accurately and efficiently. This machining process is especially ideal for cutting materials that cannot be cut by laser or thermal cut. Metallic, non-metallic and advanced composite materials of various thicknesses can be cut by this process. This process is particularly suitable for heat sensitive materials that cannot be machined by processes that produce heat while machining. The schematic of abrasive water jet cutting is shown in Figure which is similar to water jet cutting apart from some more features underneath the jewel; namely abrasive, guard and mixing tube. In this process, high velocity water exiting the jewel creates a vacuum which sucks abrasive from the abrasive line, which mixes with the water in the mixing tube to form a high velocity beam of abrasives.

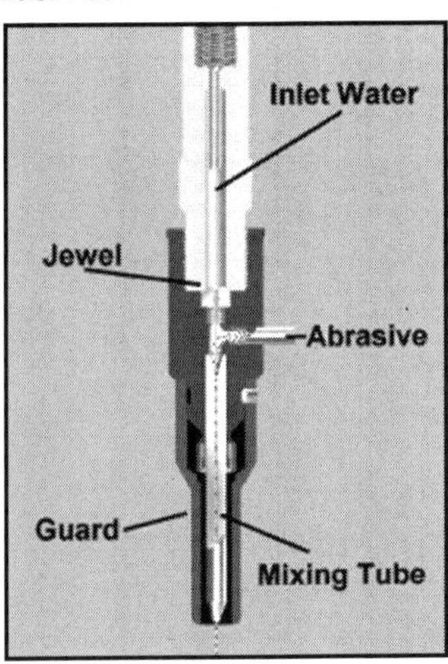

Fig. Abrasive water jet machining

APPLICATIONS

Abrasive water jet cutting is highly used in aerospace, automotive and electronics industries. In aerospace industries, parts such as titanium bodies for military aircrafts, engine components (aluminium, titanium, heat resistant alloys), aluminium body parts and interior cabin parts are made using abrasive water jet cutting. In automotive industries, parts like interior trim (head liners, trunk liners, door panels) and fibre glass body components and bumpers are made by this process. Similarly, in electronics industries, circuit boards and cable stripping are made by abrasive water jet cutting.

Advantages of abrasive water jet cutting

- In most of the cases, no secondary finishing required
- No cutter induced distortion
- Low cutting forces on workpieces
- Limited tooling requirements
- Little to no cutting burr
- Typical finish 125-250 microns
- Smaller kerf size reduces material wastages
- No heat affected zone
- Localises structural changes
- No cutter induced metal contamination
- Eliminates thermal distortion
- No slag or cutting dross
- Precise, multi plane cutting of contours, shapes, and bevels of any angle.

Limitations of abrasive water jet cutting

- Cannot drill flat bottom
- Cannot cut materials that degrades quickly with moisture
- Surface finish degrades at higher cut speeds which are frequently used for rough cutting.
- The major disadvantages of abrasive water jet cutting are high capital cost and high noise levels during operation.

Fig. Abrasive water jet cutting

Fig. Steel gear and rack cut with an abrasive water jet

A component cut by abrasive water jet cutting. As it can be seen, large parts can but cut with very narrow kerf which reduces material wastages. The complex shape part made by abrasive water jet cutting.

3

Biotechnological Medical Device Industry and Medical Manufacturing Technology

Reducing the size of surgical instruments for minimally invasive procedures is critical to limiting operating and recovery times, which has become even more important as rising health care costs push hospitals to shorten patient stays. As a result, the medical device community must deliver new and innovative designs that provide the same level of precision and function in much smaller packages.

With this demand comes the need for new and innovative ways of manufacturing miniature parts. For silicone components, this means taking the injection molding process to the microscale level. Whether producing elastomeric components that can be used independently, or overmolding silicone to more rigid materials, tiny and complex parts have helped support the success of minimally invasive surgery.

So how small is micro? Depending on the industry the answer varies. But for silicone in the medical industry, a micro part weighs around 0.001 grams or less. Parts can also include more intricate features on larger components. This increase in complexity brings new manufacturing challenges. Starting down the micromolding path is a decision that needs careful thought and consideration. A good place to start is to look at the critical areas of the process: mold design, manufacturing process, and quality.

Mold design for micromolding has become possible due to advances in machining technology. The most commonly used machining processes are electrical discharge machining (EDM) and high-speed milling. Both work well for cutting molds that meet the tight fit tolerances of micromolded parts. But each process has its advantages and disadvantages, mostly related to mold finish. Because the burning in EDM cannot be perfectly controlled, molds may be left with slight surface variations, producing a rough finish. On a macro molding level this may not be a big issue but micromolding deals with measurements of tens of thousandths of an inch, which means any finish variation is a primary concern. Milling also suffers from issues in variation because cutter length and steel hardness can cause flexure that may not be precise enough for the tolerances though the mold finishes produced by milling are smoother than those of EDM. On the other hand, EDM produces crisper vertical walls and sharp angles than can be cut with milling operations. Because the cost for both of these processes is similar, deciding which to use depends on part complexity and the time it takes to cut the mold. Generally, cutting the mold involves some combination of the two technologies. It is definitely advantageous to have tooling resources available in-house because outsourcing may cause a breakdown in the communication necessary to meet timelines.

Once the mold has been built, it is time to look at the how to manufacture the actual part and that begins with the material dosing equipment. Micromolding requires small volumes, and the silicone must be injected at a high velocity for an even flow. Short cure times and miniscule sizes force more automation because it is difficult to see parts. Also, issues with static electricity can cause component to cling to everything, even the operators. The less handling involved, the less problematic the micromolding process.

Additionally, detecting defects requires specialized micro and nanomeasuring systems with high magnifications because at this scale, standard magnifications are ineffective. These systems are generally noncontact due to size constraints and part strength, particularly with elastomers. Even how the parts are packaged must be customized to deal with such minute components. Ideally, everything would be completed in-line to reduce contamination and avoid losing parts, which is likely to happen when they are moved. As for shipping, smaller counts per package and multiple packages are highly recommended. This approach makes things easier for loading and unloading the finished product. It also mitigates the risk involved in transporting high value product, potentially thousands of parts, which fit in a few square inches.

In addition to new challenges in manufacturing, there are also new challenges in quality. Having a knowledgeable, experienced micromolding partner and open lines of communication becomes important in establishing the new quality standards. The path to full production can be problematic if the measurement methods haven't been established well in advance, especially when dealing with differences of tenths of thousandths of an inch. Even changes in simple things like the lighting on the part and where the vision camera focuses can have a huge impact on measuring small parts.

On the up side, micromolding provides new design opportunities for engineers that were previously unattainable. For example, secondary operations such as assembly can be eliminated because the number of components going into a device are reduced due to insert and overmolding capabilities. Also, micromolded part design is a bit more flexible than in standard molded parts. For example, engineers normally avoid sharp corners to eliminate defects that result from uneven material flow and shrinkage during curing. But micromolded parts can have nice sharp corners due to their high injection velocity, small size, and low shrink rates. The negligible shrink rates allow for variations in wall-thickness that are unheard of in macro scale parts. And specific to the injection molding of silicone, designers can work with undercuts they normally would not use with thermoplastics or other thermoset materials.

In the end, micromolding is becoming more and more prevalent as surgeries become less invasive. With so many patients needing better and cheaper healthcare options, it is only a matter of time before suppliers are pushed to provide smaller devices. Being prepared and understanding the complexities involved in producing tiny parts will help medical component suppliers get a head start on the next phase of manufacturing technology.

OVERVIEW OF THE BIOTECHNOLOGY AND MEDICAL DEVICE INDUSTRY

While economic uncertainties cloud the outlook for several industries, biotechnology and medical device sectors are likely to continue posting strong growth in revenues and earnings. At the heart of these sectors' positive fundamentals are favorable demographics and continued growth in health care expenditures. Two worldwide trends—the aging of the post-World War II generation and the lengthening of life expectancy—bode well for the biotechnology and medical device industries. Globally, the over-60 year-old population is expected to more than triple between 2000 and 2050, representing more than a fifth of the world's population. In the United States, the over-65 year-old population is expected to more than double between 2001 and 2030, thus expanding the market for biotechnology products and medical devices.

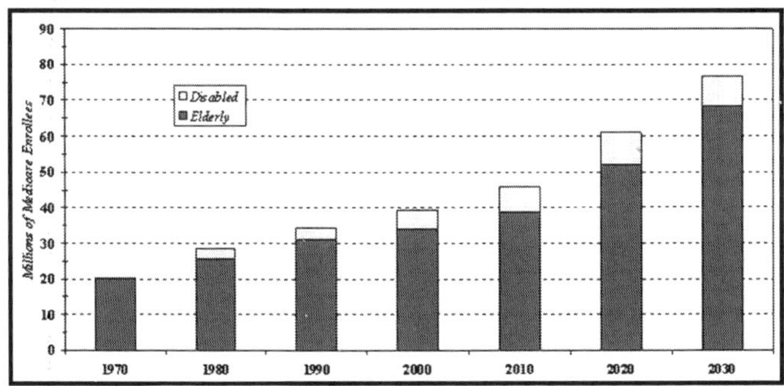

Fig. Medicare Beneficiaries, 1970-2030

Note: 2010, 2020, and 2030 is estimated.

Currently, Americans 65 years and older account for 13 percent of the nation's population yet consume an estimated 33 percent of all U.S. pharmaceutical output. Medical device companies are also well-positioned to benefit from this graying of the population, as the use of such medical devices as pacemakers, defibrillators, stents, and orthopedic implants increases in frequency with older patients. Going forward, an aging segment of baby boomers will provide further stimulus for industrywide demand for biotechnology products and medical devices. Baby-boomers, according to the U.S. Census Bureau, account for almost a quarter of the population. Much of the increase in the over-65 year segment of the U.S. population will occur after 2010 when the leading edge of the baby boom generation begins to turn 65.

Americans now have a life expectancy at birth of 76.5 years. Those Americans who have reached their 65th birthday are likely to live another 16-19 years. Although more people than ever will be living longer, they will not necessarily be free of health problems. According to the World Health Organization (WHO), the incidence of cancer, heart disease, and other chronic diseases—which currently cause some 28 million deaths worldwide annually—is expected to increase in the future.

This is largely due to unhealthy lifestyles, poor diet, obesity, smoking, and/or lack of exercise. The WHO projects a doubling of cancer cases in most countries over the next 25 years. In the United States, the top four causes of death are all disease-related: heart disease, cancer, stroke, and respiratory disease (National Center for Health Statistics, 2002). Biopharmaceuticals and medical devices should play an increasingly important role in combating the ravages of these diseases and benefiting the elderly, as well as dealing with more obscure maladies for which major pharmaceutical firms may not look to develop treatments.

Health care expenditures for hospital care, physician care, drugs, medical devices and medical nondurables rose 9.6 percent to $1.42 trillion in 2001 (Centers for Medicare & Medicaid Services, 2002). The rate of spending growth is expected to continue at an average annual rate of 6.9 percent into the next decade when health care expenditures will exceed $2.8 trillion and represent 17 percent of GDP. Both medical devices and pharmaceuticals account for a relatively small component of total U.S. healthcare expenditures at 3 percent ($19.9 billion) and 8 percent ($141.8 billion), respectively. While spending on medical devices is expected to grow at an average annual rate of 5.4 percent, spending on pharmaceuticals is expected to escalate to an average annual rate of 11 percent.

The government healthcare entitlement programs of Medicare and Medicaid also have a significant impact on biotechnology and medical device industry, particularly through changes in spending levels and reimbursement rates. The Centers for Medicare and Medicaid Services (CMS), which oversees these federal healthcare programs, is the single largest purchaser of health care in the world with aggregate outlays of $351 billion in 2001. Medicare and Medicaid indirectly fund large segments of biotechnology and medical device markets. Such funding is particularly important for makers of high-tech medical device products.

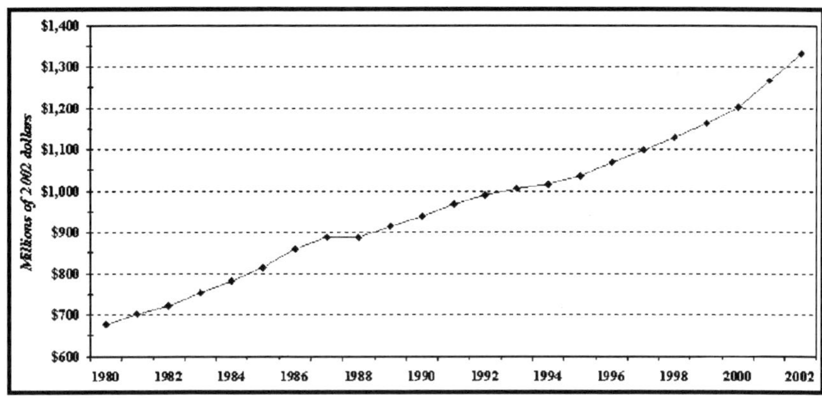

Fig. Personal Health Care Expenditures in United States, 1980 -2002

BIOTECHNOLOGY AND MEDICAL DEVICES DEFINED

Biotechnology and medical devices are distinct yet complementary sectors dynamically linked to each other within the biosciences industry. Developments within both biotechnology and medical devices are becoming more complementary over time, as medical devices of increasing sophistication and miniaturization are being used to deliver new pharmaceutical and biotechnology products.

Biotechnology

Biotechnology is unique among industries in that it is not defined by its products, but by the technologies used to make those products. Biotechnology is formerly defined as the application of biological knowledge and techniques pertaining to molecular, cellular, and genetic processes to develop products and services. Biotechnology has potential applications in a wide variety of industries— from natural resources (genetic engineering of plants and animals) to manufacturing (food processing and chemical engineering) and even computing (bio-computers). Industrywide revenues for public biotechnology firms exceeded $28 billion in 2001, with an anticipated increase to $33 billion in 2002. The biotechnology industry consists of more than 1,450 public and private companies with over 141,000 employees in the United States (Ernst & Young, 2002).

Because biotechnology is application-oriented, sales of products made through biotechnology cross several traditional industries. Data gathered concerning industries are usually collected based on the product or service produced, not on the method of production or technology used. Consequently, we do not know the full extent of sales of products made through biotechnology as a separate industry; almost all product sales are subsumed within the various traditional industry categories as pharmaceuticals, foods, chemicals, plastics, medical devices, environmental services, and research and development.

Table. Applications of Biotechnology

Healthcare	**Veterinary**	**Agriculture**
Pharmaceuticals	Genetic engineering	Diagnostics
Therapeutics	Monoclonal antibodies	Biopesticides
Vaccines	Recombinant interferons	Microbial diagnostics
Gene therapy	Transgenic animals	Environmental adaptions
Growth & other hormones		Genetically modified foods
Pharmacogenomics	Industrial biotechnology	Natural fertilizers
Diagnostics	Energy conservation	Nutraceuticals
Biosensors	Pollution reduction	Transgenics
DNA probes	Waste reduction	
Monoclonal antibodies	Bioelectronics	Forestry biotechnology
Polymerase chain reaction	Organic chemicals	Genetic engineering
Genomics/proteomics		Biological pest control
Biological instrumentation	Environmental biotechnology	Tissue culture propogation
Biological devices	Bioremediation	Transgenic plants
Drug-coated stents	Biotreatment	
Microdevices	Immunoassays	Marine biotechnology
Bone growth devices	Microbial mining	Biomedical research
Cartilage regeneration	Environmental monitoring	Food supply
Tissue replacement		Therapeutics
Wound healing devices	Food Processing	
Skin grafts & wound closures	Microbial starter cultures	
	Enzymes and vitamins	
	Food contamination tests	

Biotechnology firms are not separately classified in either the Standard Industrial Classification (SIC) system or its successor, the North American Industrial Classification System (NAICS). Instead, most biotechnology companies are assigned to one of two broader industry categories encompassing research and development and drug manufacturing, namely NAICS five-digit professional, scientific, and technical services industry 54171 (Research and development in the Physical, Engineering, and Life Sciences) or the NAICS manufacturing industry group 3254 (Pharmaceutical and medicine manufacturing).

By far, the largest category of biotechnology applications is within health and medicine—diagnosing, treating, and preventing disease. Biotechnology, however, should not be equated with medical technology or high-tech medicine. Many medical technologies fall outside the genetic and cellular manipulation of biotechnology.

In Washington State, nearly three-fourths of the 133 biotechnology companies are classified as healthcare—therapeutics, diagnostics, and genomics/informatics. The remaining one-fourth of Washington's biotechnology companies is dispersed across various other market segments of natural resources (agriculture forestry and marine), food processing and other industrial, and environmental biotechnology. Most of the 19 publicly-traded biotechnology companies in Washington are within the healthcare and medicine segments.

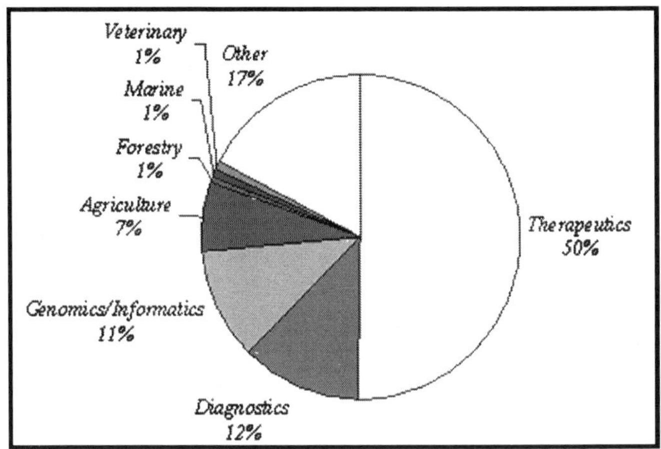

Fig. Washington Biotechnology Companies by Market Segment, 2002

Medical devices

Medical device manufacturers produce a wide range of products used for the diagnosis and treatment of ailments. These include surgical and medical instruments, electromedical and electrotherapeutic apparatuses, irradiation apparatuses, surgical appliances and supplies, ophthalmic goods, in vitro diagnostics, and laboratory equipment. As in biotechnology, most medical device companies are assigned to one of three categories encompassing medical and surgical manufacturing, namely NAICS five-digit manufacturing industry 33911 (Medical Equipment and Supplies Manufacturing), or NAICS six-digit manufacturing industries 334510 (Electromedical and Electrotherapeutic Apparatus Manufacturing) and 334517 (Irradiation Apparatus Manufacturing).

In Washington State, two-thirds of the 57 medical device companies produce either medical and surgical equipment (e.g., syringes, catheters, blood transfusion equipment, surgical clamps, and stents) or medical and surgical appliances and supplies (e.g., orthopedic devices, prosthetic appliances, surgical dressings and sutures). One-fifth of the companies are classified within the electromedical and electrotherapeutic apparatuses (e.g., magnetic resonance imaging equipment, ultrasound equipment, pacemakers and defibrillators, hearing aids, and electrocardiographs), with the remaining companies in irradiation apparatus, laboratory apparatus, or in vitro diagnostic substances.

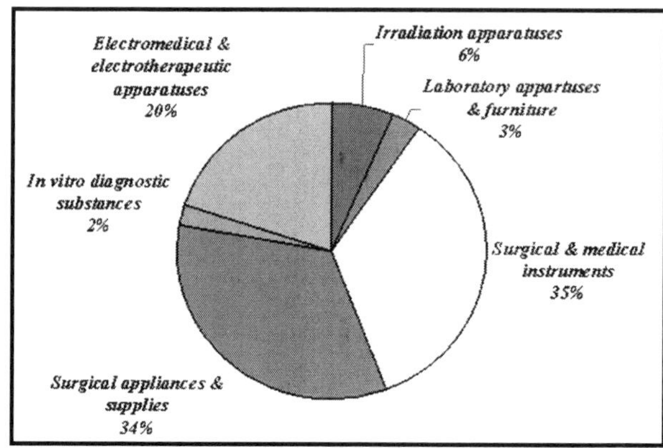

Fig. Washington Medical Device Companies by Market Segment, 2002

STRUCTURE OF THE BIOTECHNOLOGY AND MEDICAL DEVICE INDUSTRIES

The biotechnology and medical device industries have a number of important characteristics that distinguish them from each other and from other industries. Here follows a brief overview and development of these two industry sectors, their current structures, and some of the important aspects of the regulatory and competitive environment surrounding firms in each sector.

Biotechnology

The structure of the biotechnology industry seems to be in a continual state of flux. Advances in bioscience over the last three decades have led to a circular evolution of the biotechnology industry structure: from the domination of large-scale pharmaceutical firms to the entry of many small, innovative start-ups to alliances between the large and small for more efficiency (and in some cases, survival) to currently a few dominant biotechnology powerhouses.

One of the earliest major milestones for biotechnology, which led to an explosion in research and production mechanisms and a sea-change in the industry's organization, was the successful recombining of DNA by Stanley Cohen and Herbert Boyer in 1973. This advance in genetic science propelled the industry along a couple of different paths. One path employed genetic science as a process technology, that is, using the methods of Cohen and Boyer to mass produce proteins as therapeutic agents. Genetic engineering required new techniques and changes in R&D efforts by firms. Before genetic engineering, a small number of proteins could be manufactured either from natural sources or by organic chemical methods. Genetic engineering made it possible to produce large quantities of proteins, opening an entire new field in drug research. The number of these new "biotechnology" firms grew from nonexistent in the mid 1970s to over 1,450 firms less than three decades later, spawning a new industry and transforming an old industry—pharmaceuticals—in the process. Another path employed biotechnology techniques as a primary research tool for discovering and manufacturing conventional "small molecule" drugs. This trend helped reinforced the dominance of the large pharmaceutical firms, which were able to leverage their competency in various R&D arenas to build on the knowledge already codified in the academic literature.

The academic research done in the universities during the 1970s and 1980s spawned many small, innovative start-up companies, beginning with Genentech, formed by Boyer and Robert Swanson in 1976. Start-ups increased during the 1990s along with a growth in merger activity as large biotechnology companies purchased innovative start-ups. Often, the mergers occurred because target R&D firms, while rich in talent, were poor in capital and resources to commercialize their research into products. These start-ups needed the distribution and production prowess of larger firms to take their products to market. In contrast, larger firms needed new ideas but found it more economical to acquire brain-rich start-ups than to expend scarce resources on cutting-edge in-house research. Moreover, by acquiring an established firm, a larger firm was able to diminish the uncertainty inherent in R&D efforts. Thus, few firms, even extraordinarily successful ones, do not grow into large biotechnology companies. Instead, biotechnology research firms tend to sell or license their technologies to larger concerns, or form joint ventures with them, or outright sell their entire companies.

Today, there are only a handful of companies that dominate the biotechnology industry. Each of these companies' dominant position is in specific market segments, with the goal of broadening their product pipelines by acquiring smaller biotechnology concerns or partnering with undercapitalized entities. The biotechnology industry in Washington State is recently established, with the majority of companies founded since 1990. Before 1990, less than a third of all biotechnology firms had been started in Washington. At that time, much of the biotechnology–related research in the state was academic-oriented at the University of Washington and the Fred Hutchinson Cancer Research Center (founded in 1972).

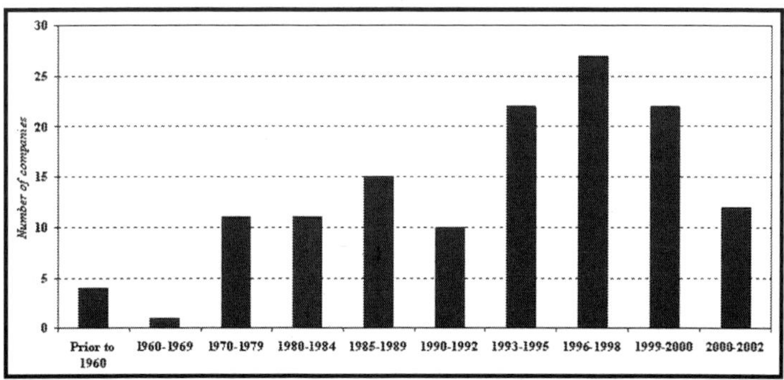

Fig. Washington Biotechnology Companies by Year Founded

Most biotechnology firms in Washington, as elsewhere, are small companies. In 2002, more than four-fifths of all biotechnology firms had less than 100 employees. For the largest biotechnology companies, three are actually non-profit research institutes or federally-funded research laboratories. In general, most biotechnology companies have added employees over the years, fueled solely by research and development activities and the promise of eventual approval of medicines-in-development. According to the latest PhRMA survey, there are currently 371 biotechnology medicines in development by 144 companies. Washington-based companies account for twenty-five of these biotechnology medicines in various stages of development. Of the 95 biotechnology medicines already approved by the Food and Drug Administration, only a few drugs have reached "blockbuster" status. One of these "blockbuster" drugs, Enbrel®, was developed by Immunex, resulting in spectacular employment growth within the firm following the initial approval and marketing in late 1998.

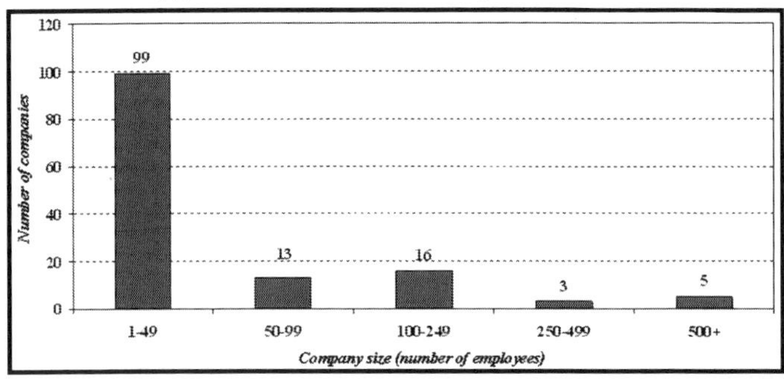

Fig. Size of Washington Biotechnology Companies, 2002

"Science is the pilot of industry" aptly describes biotechnology, for it is one of the nation's most research-intensive industries. This knowledge-based sector is largely dependent upon universities as a significant source of basic research and applied technologies. Basic bioscience research and discovery conducted at academic universities and institutes can be powerful sources of economic growth, especially in collaboration with the private sector.

A major player behind these fundamental developments within biosciences is the Federal government. A wide variety of federal agencies provide funding to biosciences and medicine and biotechnology, but by far the largest single funding agency of such research is the National Institutes of Health. The lion's share of federal health-related funding is directed toward supporting research activities at academic universities, medical schools, and non-profit institutions, underscoring their critical role within the biotechnology industry.

During the 1990s, the growth rate of NIH funding for biosciences research has been 7.8 percent annually. Total NIH spending, in fact, doubled during the decade from $6.5 billion in 1991 to over $13 billion in 2000. In 2001, NIH disbursed a total of $14.9 billion for research activities in biosciences. Among all states, Washington is ranked eighth in total NIH funding, with a per capita share far greater than the national average. In Fiscal Year 2002, Washington State universities, institutes, and companies received $649 million from the National Institutes of Health for biosciences research.

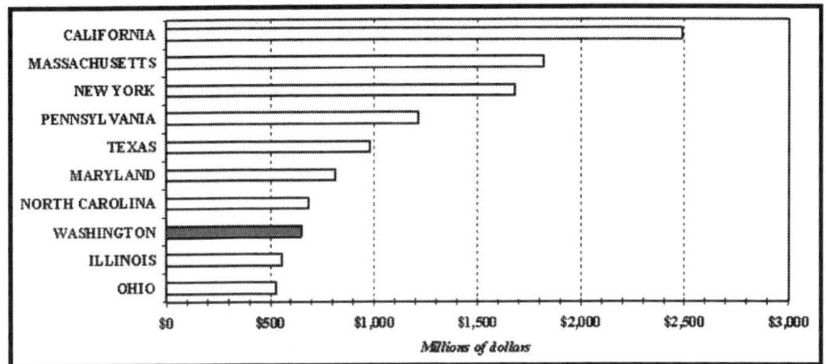

Fig. Top Ten National Institutes of Health Grantee States, FY2002

Washington's academic universities and research institutes are the vital source of the state's biosciences R&D industry. NIH funding within Washington State universities, research institutes, and companies have more than doubled in ten years. Broad, early-stage, basic research studies at the University of Washington, Fred Hutchinson Cancer Research, Washington State University and others—funded largely by the National Institutes of Health, National Science Foundation, Department of Energy, and Department of Defense—are essential to private sector innovation. In FY 2002, the National Science Foundation awarded $21.6 million for biosciences-related research to universities, research institutes, and companies in Washington. The Pacific Northwest National Laboratory's biosciences program is underwritten by U.S. Department of Energy funding; amounting to $72.6 million in FY2002.

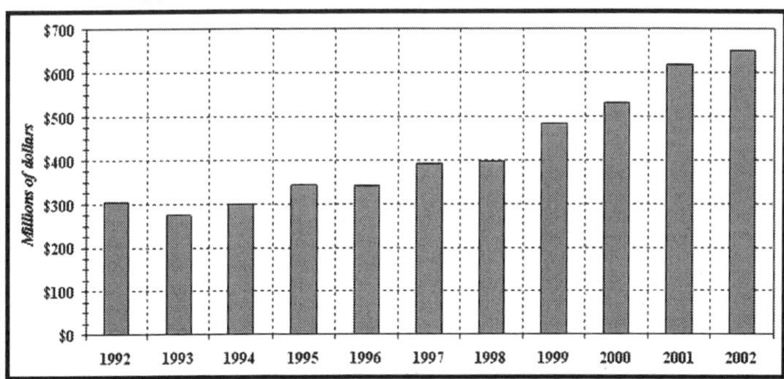

Fig. NIH Funding to Washington State, FY1992-2002

The University of Washington ranks first among all public universities (third among all universities) in NIH funding. In FY 2002, bioscience researchers at the University of Washington received a total of $389.6 million in NIH funds. The Fred Hutchinson Cancer Research Center ranked 24th among all institutions (but first among all research institutes) in NIH funding with $161 million (FY2002).

Recent studies underscore the premise that basic university science is integral to the successful commercialization of scientific discoveries. While universities and research institutes are not equipped to develop and market products, they can patent and license their inventions, and in turn transfer core technologies to biotechnology entrepreneurs. The history of biotechnology and medical devices is rather fluid—in ideas and people—between universities and institutes and companies. Most successful companies trace their origins back to academia; and many academic institutes profit from royalties on their faculties' inventions.

In Seattle, the region's biotechnology industry has been driven by spin-offs from the University of Washington and Fred Hutchinson Cancer Research Center. A majority of the biotechnology firms in the state trace their roots to technology breakthroughs developed at these two research centers. Two of the state's earliest and ultimately largest biotechnology firms—Immunex and Zymogenetics—were started in 1981 by researchers from the "Fred Hutch" and the University of Washington.

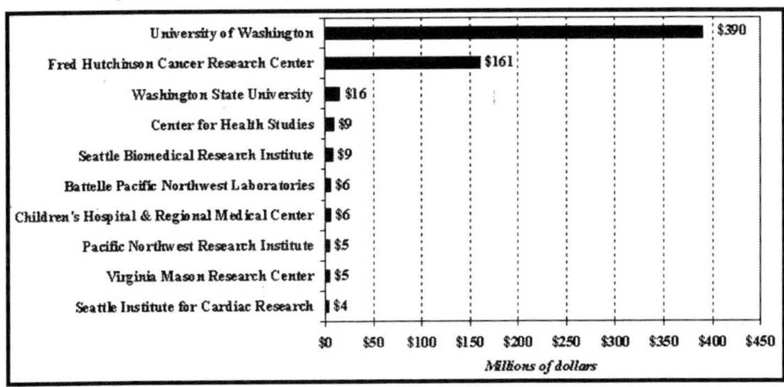

Fig. Top Ten NIH-Funded Institutions in Washington State, FY2002

Biotechnology demands a stock of innovations; indeed, growth in the biotechnology industry is inextricably linked to the rate of intellectual property generation. Predicated on knowledge creation and intellectual property, firms and institutes generally seek to patent new products and processes. Small firms in particular, seek to develop their intellectual property and sell it to larger firms for manufacture and distribution.

Patent data are classified according to product or technology characteristics (US Patent and Trademark Office, 2002). Combined with medical devices, biotechnology-related patents represent a large and growing fraction of all patents granted in the United States and Washington—about 14.3 percent and 15.7 percent respectively between 1997 and 2001. Private firms own most of these issued patents. Universities and government agencies also hold a sizeable share of patents because they sponsor a considerable amount of research. In Washington, Zymogenetics and Immunex are among the leading biotechnology patent holders with the University of Washington and the Fred Hutchinson Cancer Research Center also highly ranked. Overall, Washington State ranks 13th among all states in biotechnology patents and 10th in medical device patents.

The availability of capital plays a critical role in the growth and development of the biotechnology industry. Biotechnology requires expensive and time-consuming research and even resultant promising compounds must undergo a long process of testing and approval. Numerous biotechnology research projects and promising ideas fail to materialize revenue. Even those firms that ultimately succeed record many years of losses during research and development. As a result, large amounts of financial capital over an extended period are required in order to grow and develop firms within the biotechnology industry. Various financial capital programs are accessed by biotechnology companies—Small Business Innovation Research (SBIR) grants, venture capital, research alliances, and initial public offerings. These programs represent different phases in the life cycle of firms and in product development.

For early-stage financing for development and commercialization, many biotechnology and medical device research institutes and companies access the Small Business Innovation Research (SBIR) and Small Business Technology Transfer Research (STTR) programs. Washington State research institutes and companies have been increasingly successful in obtaining financial support from these competitive programs; in FY2002 alone, 50 research institutes and companies in Washington State were awarded grants totaling $23.5 million.

By far, the most important source of start-up capital is organized venture capital—private investments made by professional fund managers that typically specialize in a set of technologies. Venture capital investment finances most biotechnology companies from their inception through the early years of research and development needed to prove the potential of a promising idea. A firm may receive several rounds of financing as it develops its biotechnology products. Due to the considerable expense and long lead times associated with product development, venture capital essential to the start-up of firms may go several years before generating revenues.

Venture capital, as such, is a good leading indicator for the biotechnology industry. Since 1995, biotechnology firms in Washington have attracted over $764 million (77 deals) in venture capital investment. In 2000 during a capital market boom, biotechnology firms attracted nearly $3 billion in venture capital investments, $281 million of which was invested in Washington biotechnology companies.

Workforce

Biotechnology requires a highly specialized labor force. In addition to the growth patterns displayed by the biotechnology and medical device sector, an understanding of the occupational structure of the industry is important for the constructing of sound economic development strategies as well as the design of employment and training programs. What are the important occupations associated with the biotechnology sector? In other words, what is the occupational mix for biotechnology? Second, how important is biotechnology as a source of employment for specific occupations within the statewide labor market? The occupational distribution of Washington biotechnology workers is weighted heavily toward executive, administrative, managerial, professional specialty, and technical occupations. The leading occupations in the biotechnology sector for 2001 are based on information available from the Washington State Employment Security Department, Labor Market and Economic Analysis Branch. The unique characteristics of biotechnology are clearly evident as there is a heavy reliance on science, medical, and technical positions. More specifically, eight of the top fifteen occupations are associated with either a science, medical, or technical skills set. Together with the other seven occupations, primarily in office-related occupations, these fifteen occupations account for 42 percent of total employment in the biotechnology sector.

Table. Leading Occupations in the Biotechnology Sector, 2001

Rank	OES	Occupation	Employment	Annual Wage	%
1	43-4111	Interviewers, Ex. Eligibility and Loan	1,505	$15,456	7.5%
2	19-4021	Biological Technicians	1,325	$26,109	6.6%
3	19-1042	Medical Scientists, Ex. Epidemiologists	767	$34,313	3.8%
4	19-4031	Chemical Technicians	763	$29,053	3.8%
5	19-2031	Chemists	547	$48,240	2.7%
6	11-9121	Natural Sciences Managers	439	$55,542	2.2%
7	13-1199	Business Operations Specialists, All Other	422	$34,577	2.1%
8	41-9099	Sales & Related Workers, All Other	406	$14,800	2.0%
9	11-1021	General & Operations Managers	350	$65,763	1.8%
10	47-4041	Hazardous Materials Removal Workers	315	$19,707	1.6%
11	29-1111	Registered Nurses	307	$51,119	1.5%
12	43-9061	Office Clerks, General	303	$19,123	1.5%
13	19-3022	Survey Researchers	297	$14,548	1.5%
14	43-6011	Executive Secretaries & Admin. Assistants	283	$30,623	1.4%
15	43-9111	Statistical Assistants	272	$24,888	1.4%
		Top Fifteen occupations	8,301	NA	41.6%

Notes:

NA is not available; Annual wage is for entry workers, with no experience.

While occupational mix provides some indication of the importance of which occupations are most important within biotechnology, it is also useful to examine the relative importance of biotechnology as a source of employment for various occupations. Table reports biotechnology's share of total employment within these same major occupations. While the industry as a whole accounts for 0.8 percent of total state employment, it can claim a much higher share of employment in particular occupations. The majority of survey researchers—used in clinical trials, chemists, and chemical technicians—are mostly found in biotechnology.

Table. Occupational Concentrations in Biotechnology, 2001.

			Biotec h	Education
OES	**Occupation**	**Employment**	%	**Required**
19-3022	Survey Researchers	502	59.2%	3
19-2031	Chemists	974	56.2%	1
19-4031	Chemical Technicians	1,395	54.7%	2
19-4021	Biological Technicians	2,933	45.2%	2
43-4111	Interviewers, Ex. Eligibility and Loan	3,522	42.7%	2
19-1042	Medical Scientists, Ex. Epidemiologists	1,957	39.2%	2
11-9121	Natural Sciences Managers	1,285	34.2%	2
43-9111	Statistical Assistants	1,184	23.0%	2
47-4041	Hazardous Materials Removal Workers	2,439	12.9%	4
13-1199	Business Operations Specialists, All Other	28,247	1.5%	2
43-6011	Executive Secretaries & Admin. Assistants	22,349	1.3%	3
11-1021	General & Operations Managers	30,909	1.1%	2
41-9099	Sales & Related Workers, All Other	40,000	1.0%	3
29-1111	Registered Nurses	41,912	0.7%	2
43-9061	Office Clerks, General	59,070	0.5%	3

Notes:

Education required: 1—long (4 years or more); 2—moderate (1-4 years); 3—short (1-12 months); and 4—little (less than 1 month).

Most of these occupations are projected to increase faster than overall statewide employment during the 2000-2010 time period. Whether or not this demand will be satisfied will largely be a function of the extent to which the biotechnology industry faces competition from other sectors for these occupations and the degree to which individuals are educated and trained either within or outside of Washington State. Each of these occupations has educational codes indicating the level of training and education required. In this table, relatively few occupations require "long" or "little" preparation. Most of these leading occupations require some formal education/training that is closely related to which a community and technical college or four-year college degree program provides. A consortium of local-area schools and colleges in Washington provide targeted training for employment in the biotechnology and biomedical fields.

MEDICAL DEVICES

The national medical device manufacturing industry generated $31 billion in sales in 2001. Medical devices are used in life-saving and life-enhancing medical procedures and include important value-added products such as infusion and related intravenous pacemakers, coronary stents, hip replacements, catheters, and implantable defibrillators.

Similar to biotechnology, the U.S. medical device industry is dominated by a few large companies that book over half of the industry's annual sales. Nationally, 80 percent of all medical device companies have less than 50 employees. Washington's medical device industry has a similar firm structure.

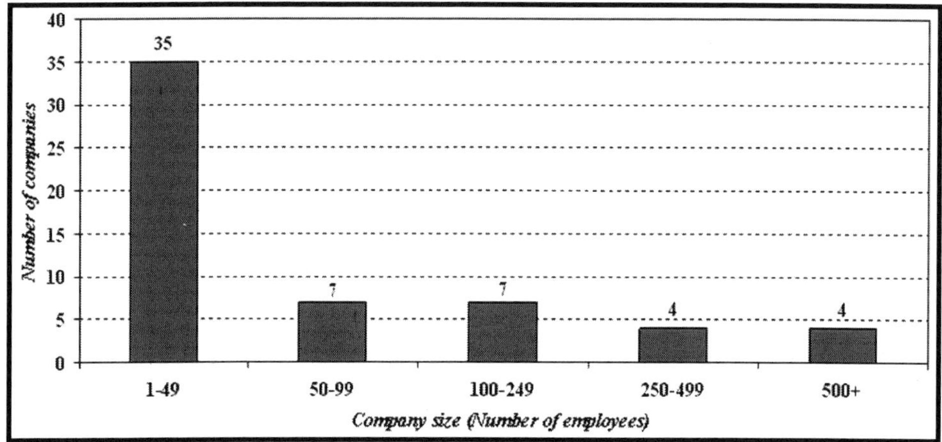

Fig. Size of Washington Medical Device Companies, 2002

These small and emerging companies historically have made a critical contribution to innovation and early development of many novel devices. Often these small companies collaborate with larger companies to bring their products to market. Some small companies try to self-market new products, but there are significant barriers to entry including funding of research and development, manufacturing, and distribution.

Unlike biotechnology, medical device product life cycles are relatively short because device makers are continually developing smaller, faster, and cheaper improvements of existing devices. In addition, patent protection of a new technology can often be challenged or circumvented, with no analogous patent to a biotechnology company's "composition of matter" patent on the compound itself.

As in biotechnology, new products are the engine of growth in the medical device industry. Fueled by aggressive R&D spending and increasing investment in new medical technologies, a plethora of new sophisticated medical devices have come on the market in recent decades. Smaller start-up companies originally developed many of these new products. In Washington State, two-thirds of the medical device companies were started since 1990.

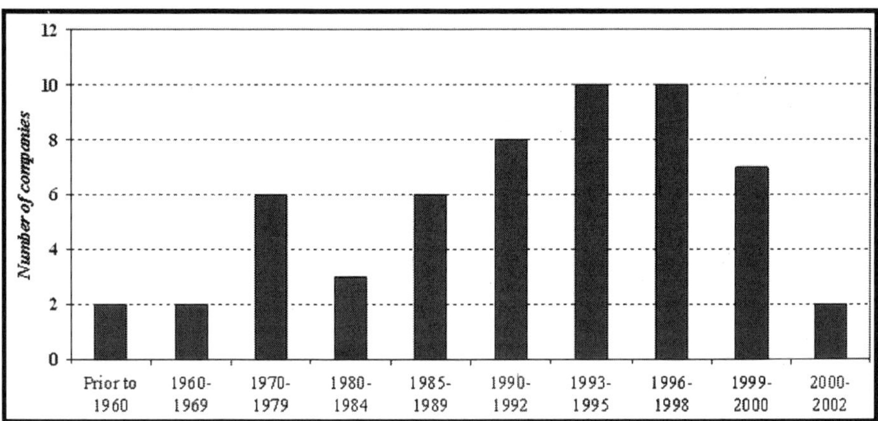

Fig. Washington Medical Device Companies by Year Founded

Medical device firms greatly benefit from basic biosciences research funded by the National Institutes of Health and other federal agencies. Collaboration between universities and medical device companies continues to increase with technology transfer, sponsored research, and license agreements. One of the early significant events within the Washington medical device industry was a 1974 technology transfer agreement reached between the University of Washington and Advanced Technology Labs, laying the groundwork for the rise of the ultrasound diagnostics segment in Washington State.

Private medical device companies fund the majority of research and development costs and rely on strong intellectual property rights to protect this research investment. As in biotechnology, any new device requires clinical studies to show the device is safe and effective, which are then submitted to the Federal Drug Administration (FDA) for review. If the device obtains FDA approval, the medical device company will generally seek Medicare coverage and payment from the Centers for Medicare & Medicaid Services. The speed of technology adoption often depends on a combination of clinical benefit data, regulatory decisions, and distribution.

The U.S. medical device industry is the recognized global leader. U.S. products account for close to half of the world's medical device market, and leading U.S. manufacturers often generate roughly half of their total sales abroad. In recent years, export growth has benefited from the development of increasingly sophisticated medical devices and an increasing emphasis by foreign governments to improve the quality of healthcare for their citizens.

The high volume of Washington medical device exports is a major contributor to the state's economy. The value of Washington medical device exports of $420 million in 2001 surged 21 percent over 1999, outpacing export growth nationwide. Washington biotechnology exports are relatively modest at $8 million. Most of the medical device exports leaving Washington are bound for Europe, North America (Canada and Mexico) or Asia. In 2001, these regional markets accounted for over 92 percent of all medical device (and biotechnology) exports from Washington.

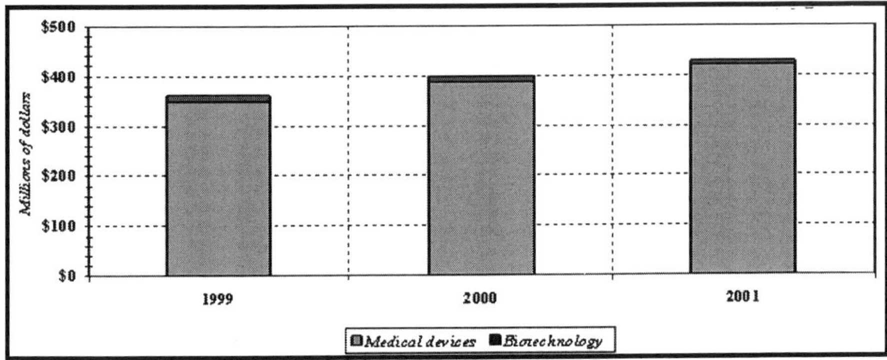

Fig. Recent Trends in Washington State's Biotechnology and Medical Devices Exports, 1999-2001

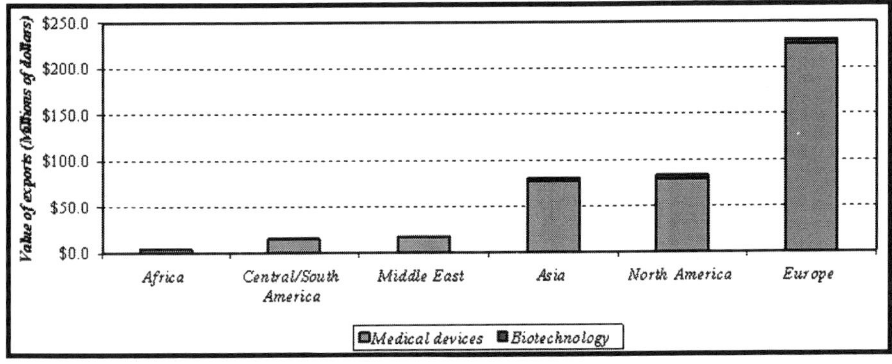

Fig. Value of Washington's Biotechnology & Medical Device Exports by Region of the World, 2001

Notes:

Central/South America includes Caribbean; Europe includes former Soviet republics; Asia includes Australia

OUTLOOK FOR BIOTECHNOLOGY AND MEDICAL DEVICE INDUSTRY

Washington biotechnology and medical devices form a large part of the state's vibrant biosciences sector. Washington is one of the leading states in biotechnology and medical devices, providing good jobs that employ highly-paid research scientists, engineers, and production workers. Through its links with other sectors, biotechnology and medical devices comprises an important part of Washington's high-technology economic base. One future scenario for this sector is one of continued trend growth, supported by growing worldwide demand for health services and the state's comparative advantages in research and discovering new technologies. For such a scenario, Washington expects increased employment in biotechnology and medical devices with an annual growth rate of 6.4 percent over the next few years. By 2005, total employment in the biotechnology and medical device sector will exceed 23,000 workers. Such sterling industry growth rates under this scenario dwarf those forecasted for statewide employment. Between 2002 and 2005, the Washington State Forecast Council projects that state jobs will grow by only 1.5 percent annually.

Whether Washington realizes this robust growth sector remains an open question. In order to ensure the future success of the biotechnology and medical devices sector—and the state's economy as a whole—Washington public policy should focus on increasing support for the industry, facilitating collaboration between industry and universities research institutes, and promoting Washington as a place to conduct business.

With continued passive state involvement, however, a counter scenario with limited industrial growth prospects is increasingly plausible. Indeed, some view the biotechnology and medical device industry to be at a critical juncture in its history. Once upbeat with rosy forecasts, industry analysts are less sanguine about biotechnology and medical device growth prospects from earlier in the year. Medical devices—with shorter product approval spans—will continue to grow and keep the overall sector buoyant. Biotechnology companies, however, face a more daunting future. Consolidation will quicken apace within biotechnology and lack of early- and second-stage financial capital for product research and development will result in further job losses for private biotechnology firms in Washington. A more proactive public sector in Washington would prove to be crucial in propelling the biotechnology and medical device industry forward.

CONTINUOUS PROCESSING IN PHARMACEUTICAL MANUFACTURING

Pharmaceutical processing in the manufacturing environment is synonymous with batch processing in the sense that each unit dosage form is identified by a unique batch. This simplified tried and true approach has been used for decades as it served well for both the industry and the regulatory bodies.

In comparison, other industries that also produce and process materials, such as petrochemical, chemical, polymer, food etc., have steadily moved to continuous processing technologies in manufacturing, driven mainly by cost and quality considerations. A recent article comparing batch vs continuous processing discussed some examples of the reason, other than tradition, why the pharmaceutical industry is dominated by batch processing. The lack of flexibility in batch processing to respond to increasing levels of growth was cited as the primary driver for why other industries have moved to continuous processing technologies. Other goals that influence the decision to move from batch to continuous processing included the desire to minimize the required size of new manufacturing plants, as well as the need to efficiently use the available capacity. Recently, both the pharmaceutical industry and the FDA agreed that an overhaul of the manufacturing regulations that apply to innovative processing methods will prove beneficial for the patient that both of them serve. A highlight of this was illustrated in a recent article in Wall Street Journal, which provided a description of how the industry and the FDA are working together in several joint initiatives to apply new quality testing methodologies.

Historically, pharmaceutical companies have competed solely on the basis of innovation through new drugs for medical needs. A recent review of drug development costs stated that capitalizing out-of pocket costs to the point of marketing approval yields a total pre-approval cost estimate of US $802 million (2000 dollars) per new drug.

In a best case scenario, the R&D costs can be expected to remain the same or increase slightly in the future. When combined with other factors, such as increases in competition, further increases in proportion of generic utilization, opening of new markets, and the socioeconomic pressures for price controls, it is evident that the industry has to look for other ways to reduce costs. Currently, new technologies and techniques, including proteomics, genomics, and the use of biomarkers, appear to be creating a future where blockbusters, as we currently define them, may or may not exist. The future instead will consist of many "customized" small volume drugs that take into consideration a patient's specific subcategory of disease and genetic makeup.

These overall shifts will translate into manufacturing many more new products. When all of the above factors are summarized, the same cost and quality drivers that have affected other industries are forcing the pharmaceutical industry to look for ways to improve quality while maintaining or reducing manufacturing costs, which today account for 36% of the industry's cost. Continuous processing technologies provide one possible path forward for the industry to reduce the cost of manufacturing, with the objective to convert selected unit operations and processes from batch to continuous mode along with appropriate real time characterization using state of the art process analytical technologies.

REGULATORY ASPECTS

The FDA regulatory definition of batch is:

A specific quantity of a drug or other material that is intended to have uniform character and quality, within specified limits, and is produced according to a single manufacturing order during the same cycle of manufacture. It appears, therefore, that regulatory definitions are already in place to support the concept of a period of time, being a "batch" for the sake of tracking and quality assurance. This interpretation, if accepted, would assist in moving to continuous processing which is by definition a single cycle of manufacture.

The overall issue of new technology introduction into the pharmaceutical manufacturing area has been very restrained, however this is changing quite rapidly from the regulatory perspective. The FDA has recently issued a draft guidance to the pharmaceutical industry in its; Guidance for Industry "PAT ¬ A Framework for Innovative Pharmaceutical Manufacturing and Quality Assurance." The goal of this guidance is to describe a regulatory framework on which industry and government can together increase the level of innovative pharmaceutical manufacturing technologies by the removal of actual and perceived barriers. The draft guidance document states, "Process Analytical Technology, or PAT, should help manufacturers develop and implement new efficient tools for use during pharmaceutical development, manufacturing, and quality assurance while maintaining or improving the current level of product quality assurance."

The background of this guidance is centered around the concept that while conventional pharmaceutical manufacturing is accomplished using batch mode, new opportunities exist to improve the efficiency and quality of the pharmaceutical manufacturing process.

This is an attempt to introduce 21st century technology into the pharmaceutical industry to better respond to the rapidly changing marketplace for ethical pharmaceutical products. The utilizing of new approaches to pharmaceutical manufacturing, while maintaining the concept that quality cannot be tested into a product, but must be built in by design, leads to the concept of continuous processing. In specific, the draft guidance document states, "Facilitating continuous processing to improve efficiency and manage variability." These regulatory statements should further encourage pharmaceutical manufacturers to begin to exploit the benefits of continuous processing.

DRUG SUBSTANCE MANUFACTURE

The manufacture of drug substance, or API (active pharmaceutical ingredient), involves several unit operations/processes. Typically, it involves several stages of reactions in which different functional groups are attached to the starting raw material. The products formed after each stage of reaction are termed as intermediates. In many cases, some downstream processing of the reaction mixture such as filtration, distillation etc. is also conducted prior to the next reaction step.

The final reaction mixture, also termed as the mother liquor, goes through multiple steps of downstream processing to produce the desired active in solid form. These steps almost always include filtration, distillation, precipitation (reactive crystallization), crystallization, drying and milling. Figure illustrates this in a schematic form. Of these, milling is inherently continuous in nature. Also, filtration and distillation can be made to operate in continuous mode without much difficulty. For filtration, two equivalent filtration units can be operated alternatively to achieve continuous operation. Once a set pressure drop is reached, the feed stream is diverted to the stand by unit, while the first unit is serviced. Continuous distillation is the norm in crude petroleum and most commodity chemical/fine chemical production. That leaves precipitation (reactive crystallization), crystallization and drying.

Continuous Chemical Reactions

Advances in reactor design and particularly in the area of microreactors over the last decade has allowed a leap in research efforts involving continuous chemical reactions. For example, a two stage continuous process was recently reported for commercial manufacture of statin intermediates that are used in the manufacture of atorvastatin (Lipitor™), currently the single largest revenue generating ethical product on the market. Researchers from the same company have published their work regarding continuous processing for generating between 50 and 60 tons per year of diazomethane, while maintaining the inventory of this highly reactive gas at less than 80g. The diazo-methane production unit is part of an integrated multistage continuous process that produces key intermediates for the latest generation of HIV protease inhibitor drugs.

Continuous stirred tank reactors and plug flow reactors have been developed and successfully used for many years. Tubular reactors, loop reactors and recent advances culminating in several designs of microreactors have enabled the researchers to have the appropriate tools to conduct continuous reactions. One such reactor is the spinning tube-in-tube (STT) system being developed. A key feature of the STT reactor's design is being able to precisely control the fluid dynamics of the reaction stream to achieve nearly instantaneous and complete molecular scale mixing of the reactants. The annular gap between the spinning and the stationary tube is reduced to less than 0.25 mm that helps to convert a volume based flow to an area based flow. Another example of a successful microreactor design is an exchangeable microreactor. The design of this reactor makes it very easy to incorporate it into an integrated system. Each individual reactor is the size of videotape with a hold up volume of 1.8ml. The high surface to volume ratio provides excellent heat transfer and improved mixing as in the previous example of STT reactor. Using the microreactor design, researchers have published their work on successful multi step synthesis of ciprofloxacin. Yet another example of a small continuous reactor design is that of a spinning disk reactor. The spinning disk reactor (SDR) has been found to be a very suitable alternative to conventional stirred tank reactors, especially for reactions involving intrinsically fast kinetics.

Continuous processes can avoid scale-up difficulties for many reactions. Prime candidate reactions that are either highly exothermic (such as nitration) or in general characterized by faster kinetics with reactive intermediates that can degrade under extended batch processing. Also, reactions that require tight control over temperature, pH or other process conditions can benefit by continuous processing. Hydrogenation is another example of a fast, frequently used reaction in bulk pharmaceutical and fine chemical synthesis that can be easily run in a continuous fashion. Overall, the key benefits of continuous reactions include better process control, enhanced margins of safety, increased productivity, and improved quality and yields.

Crystallization

Crystallization is the final step of recovering the active in the desired morphological form. Normally, the only remaining unit operations following crystallization are that of drying (to remove entrained solvent) and milling (to achieve the desired particle size distribution).

Batch crystallization is widely practiced in pharmaceutical processing, and it is safe to say that it is here to stay as the preferred process, at least for the near future. Advantages of the batch crystallization process include: simplicity of equipment, the ability to clean completely between batches, and years of operating experience. Continuous crystallizers on the other hand, have the built-in flexibility for control of temperature, supersaturation, nucleation, crystal growth and all the other process parameters that influence crystal size distribution. Also, in some cases, it has advantages when batch cooling is unacceptable, particularly where mixtures of polymorphs are formed. Semi-continuous crystallization processes often combine the best features of both batch and continuous operation, and deserve definite attention. An example of continuous operation is with crystallizers employed in a linked series of well mixed vessels, with the magma flowing from one stage to another. This strategy divides the overall temperature gradient into several stages and operates each succeeding stage at a lower temperature.

A recent technological development in the field of reactive crystallization extends the impinging jet technology to continuous reactive precipitation. The key to this technology is that it provides mixing times faster than the reaction and nucleation. This results in very high uniform supersaturation conditions, so that the end result is formation of small particles with a narrow size distribution. The spinning disk reactor, SDR, discussed in the continuous reaction section, has also been employed successfully to achieve continuous crystallization of active pharmaceutical with fast kinetics.

Drying

Historically, the most common method of drying API's has been batch drying either by tray or fluid bed dryers, with a large abundance of it by tray drying. For tray drying, heated dry air flowing through the chamber provides the driving force for the evaporation of the solvent. Tray drying has the following disadvantages: large floor space required, high labor costs associated with loading/unloading, and long drying times. In fluid bed dryers, drying is accomplished by suspending the particles to be dried directly in a stream of heated air or other gas media. Another example of batch drying includes the use of microwave energy, primarily coupled with vacuum.

Spray and drum driers are examples of driers that operate by continuous processing principles and have been used for drying of pharmaceutical actives. In spray drying, the solution or suspension is sprayed into a hot air/gas stream and circulated through the chamber. Drum driers consist of one or two slowly rotating steam heated cylinders. These are coated with solution or slurry, and drying takes place by evaporation with the final dried material being collected from the drum.

DRUG PRODUCT MANUFACTURE

The general process involved in the manufacture of drug products consists of a series of unit operations, each intended to modulate certain properties of the material being processed. Several of these commercially used unit operations are already continuous by design. For example, tableting is commercially used in unattended operation "lights out mode" and is a continuous compaction operation, run in batch mode. Milling is another common unit operation where the equipment operates in a continuous fashion but is utilized in batch mode.

Essentially, any equipment that operates on a first in/first out principle can be considered continuous by design. Equipment that operates in a continuous manner has the issue of start-up and shutdown, but operates at a steady state for the great majority of the processing time. Materials processed through such equipment experience the same level of energy input, regardless of batch size. From the standpoint of unit operations involved as practiced today, there are some that are inherently continuous in nature while there are others that are conducted in batch mode.

One of the advantages of continuous processing equipment is that the scale, or physical size, of the equipment does not change anywhere near the magnitude that batch equipment changes with increasing scale. Batch manufacture involves the changing of the scale of the equipment as batch size increases.

Usually, dramatic changes in equipment surface area to volume occur during scale-up, leading to significant differences in what the product experiences in the manufacturing vs research environments. For example, in blending with V-blenders or bin blenders, research size equipment may be 1-2 feet tall, while a production size V-blender can be 1-2 stories high. Additionally, as bin blenders are increased in capacity, they usually only get taller as the manufacturing plants generally have only 1 holder for all size bins. In another example, for suspension manufacture, the mixing tank size increases and the residence time of the material being processed also changes with scale. A laboratory size mixing tank may have heat and shear distribution kinetics so that the "time in the tank" of the product may be in the order of minutes, while a manufacturing scale batch mixing tank will require the same materials to be in the tank for hours to achieve the same endpoint.

This residence time can be a major issue for chemical and physical degradation, as well as raising potential microbial concerns. Overall, equipment designed for continuous operation is much smaller than its batch counterpart in order to process an equivalent amount of product. This difference is often translated into two or more orders of magnitude in size. A roller compactor is a good example of equipment engineered to be continuous in design, as the size of the rollers for a manufacturing size roller compactor are only somewhat larger than that of a laboratory scale system. Additionally, continuous processing equipment operates for the majority of the time at a steady state, thus easily lending itself to automation and process monitoring via PAT.

The manufacture of solid oral dosage forms can be broken into three major methodologies. The simplest one of them, direct compression, involves blending with excipients followed by tableting. A continuous direct compression system could be envisioned as several individual powder feeders that introduce the materials into a continuous blender, i.e. ribbon blender. The last section of this process would be feeding of the blended powder to a tablet press. Some activity is already ongoing in this area. Next, slightly more involved, is the dry granulation process. Here the active and selected excipients are blended and processed through a roller compactor or slugging equipment, followed by a mill. The milled material is blended with suitable excipients and tableted. The most involved, and the most common, situation includes wet granulation. Figure shows the unit operations involved when wet granulation is required for making a solid oral dosage form.

To conceive a solid oral dosage form process that is continuous, it would be necessary to conduct wet granulation, drying, milling, blending and tablet coating, in a continuous manner. Of these, milling, blending, wet granulation and drying have been successfully done in continuous mode. Continuous coating has been performed in food, flavor, and nutraceutical processing but there are no published examples of the technique being utilized in the manufacture of ethical pharmaceutical products.

Continuous Wet Granulation and Drying

For pharmaceutical processing, the early accounts of this approach, in concept, were published in two separate articles in mid 1980's. Koblitz and Ehrhardt reported on continuous wet granulation and drying. The article focused on continuous variable frequency fluid bed drying, but gave no details on granulation aspects. Berkovitch in a Manufacturing Chemist article quoted some researchers presenting these concepts in a symposium. Continuous processing of pharmaceuticals including a process for solid oral dosage form manufacturing was also discussed by Kawamura. Since then, several articles have been published over the last two decades where semi-continuous and continuous wet granulation techniques have been discussed.

Semi-Continuous Wet Granulation and Drying

A multi cell system has been recently introduced that falls in this scheme of operation. Leuenberger [23] has written several articles on functional aspects of GMC and overall advantages of continuous processing. Figure shows a schematic diagram of the system comprising of a high shear granulator followed by three stages of fluid bed drying. In the commercial scale system, the granulator is charged with 7-10 Kg of the powder blend. After granulation and wet milling, the material is conveyed sequentially through three stages of drying. In this way, four small batches (one in granulator and three in drying) are processed simultaneously and the cycle repeats for semi-continuous operation.

Continuous Fluidized Bed Wet Granulation and Drying

Continuous fluid bed systems have five or more functional zones. These are product in-feed zone, product mixing and preheating zone, spraying zone, drying and cooling zone and discharge zone. These have been reviewed in detail elsewhere [24].

Continuous Granulation Using Iverson Mixer

In this technique, powders and liquid are metered into a narrow space at the periphery of the grooved disc, which rotates at high speed. For detailed accounts, the reader is referred to the following articles. Lindberg [25] used it for studying wet granulation of placebo as well as active formulations, whereas Applegren and co-workers [26] used it for studying continuous melt granulation.

Continuous Wet Granulation and Drying Using a Planetary Extruder and Microwave Energy

A system is currently available, which uses a planetary extruder to granulate and a microwave tube through which the granulation is dried in a continuous manner.

Continuous Wet Granulation Using Twin Screw Mixer

A twin screw mixer is a modified twin screw extruder for conducting wet granulation. The process uses twin intermeshing screws that convey, mix, wet granulate and wet mill the powder blend. They offer several advantages over other wet granulation processes, and the modular nature of screw elements and a large variety available provide the user with tremendous flexibility. Detailed accounts are available in the literature [27]. Twin screw extruders themselves have also been utilized for wet granulation since the 1980's [28-30].

Continuous Drying Using Radiofrequency Energy

Both microwave and radiofrequency are electromagnetic forms of energy, commonly referred to as the dielectric energy. Microwave heating in combination with vacuum has been used extensively for drying in pharmaceutical processing [31,32]. However, until recently, radiofrequency heating has been used mainly in other industries such as food, paper, ceramic etc. Jones and Rowley [33] have reviewed several applications for drying where dielectric heating is used by itself or in combination with other methods. Ghebre-Sellassie et al. [34] have disclosed a continuous wet granulation and drying system that combines twin screw mixer (for wet granulation and wet milling) with radiofrequency energy (for drying).

OTHER CONTINUOUS PROCESSING AREAS

Other oral dosage forms including capsule filling are processed by unit operations that are intrinsically continuous, and a continuous encapsulation process, hard or soft-gel, could be envisioned in a process similar to the one previously described for direct compression of tablets. Some very interesting concepts on continuous or semi-continuous lyophilization technology were described by Rey [35].

The author looked at the food industry where continuous freeze drying is used and described a vision of what a pharmaceutical continuous freeze dryer may look like. Continuous processing concepts have also been implemented in the area of sterilization, solution manufacture, and cell culture. While no biopharmaceutical products, to our knowledge, are industrially produced by true continuous processing, several do utilize perfusion culture which can run for weeks to months, and process optimization of fed-batch fermentation has been shown to improve efficiency via continuous feeding of inducers [36]. Lastly, packaging equipment has been designed to be continuous in operation, and it is routinely used in this manner in other industries as the last section of a full continuous operation from individual starting materials until final "ready to ship" carton.

DESIGN OF MODIFIED MEDICAL IMPLANTS BY COVERED MANUFACTURING

Layered Manufacturing is a technique of fabricating parts by additive method. A 3D model generated in a CAD system is sliced into 2D profile by the software in the LM machine. The sliced-layers of the model are then added one layer at a time onto the build platform by the LM machine until a 3D part is produced. Main LM technologies used are stereolithography (SLA), laser sintering such as Selective Laser Sintering (SLS), Direct Metal Laser Sintering (DMLS) and Selective Laser Melting (SLM); Fused Deposition Model (FDM); 3D ink jet printing (3DP) techniques such as Sanders ModelMaker™, Z-Corp Ink Jet System™, 3DS Multi-Jet Modeling™, and 3DObject PolyJet™; and electron beam melting (EBM).

LM is largely being used to produce prototypes and functional parts in the engineering and manufacturing industries. Since it inception in 1986, numerous engineering and manufacturing applications utilizing LM have been documented and researched. In LM technologies, various applications in fields which are not traditionally associated with engineering and manufacturing have opened up such as in architectural and medical modeling, artistic creation [10, 36, 48] and historical restoration work [4].

The main applications are in the evaluation, visualization, validation, form fitting and functional testing in the early stage of product development process as well as tooling aids (refer to Fig 1) [39]. Only about 12% are in the Rapid Manufacturing application with growth expected in its application for customised parts. This will be particularly relevant to fields such as the medical and healthcare industry.

The principle pulling point of utilizing LM for fabrication and manufacturing parts in this field is that LM systems can produce parts of almost any geometrical complexity with relatively minimal tooling cost and time as well a significant reduction in requirements of technical expertise. The removal of tooling will reduce the cost at early stages of the product development process and avoid the lead times imposed by tooling. Minor or substantial changes to part geometry during the course of design will not incur the times and costs of producing new tooling [11, 26].

LM TECHNOLOGIES IN MEDICAL APPLICATIONS

Numerous researches have documented the use of LM in medical applications. In general LM technologies for medical applications can be categorized as follows:-

- visualisation and surgical planning
- customized orthoses and protheses implant/replacement
- Scaffoldings and tissue engineering
- Drug delivery and micron-scale medical devices

Visualisation and surgical planning

LM systems can be used to produce physical models of parts of the human anatomy and biological structures to assist in surgery planning or testing as well as for communication. 3DP system can be used to produce coloured medical models to enhance learning for students in the classroom as well as for researchers. It can be used to better illustrate the anatomy, allow viewing of internal structures and much better understanding of some problems or procedures [11, 26].

The possibility to mark different structures in different colors in a 3D physical model not only well suited for teaching purpose but can be very useful for surgery planning. The medical models generated using LM systems plays a vital role in providing tactile interaction for the surgeon with patient anatomy prior to the operation.

This facilitates the surgeon in planning and performing complex pre-operative surgical procedures and simulations. When using medical models, the surgeon will have a chance to study the bony structures of the patient before the surgery, to increase surgical precision, to decrease time of procedures and risk during surgery as well as costs and also to predict possible problem that may arise during operation [11]. As compared to conventional MRI assessment, medical models allow a more acceptable judgement over the feasibility of diagnosis and procedural planning [1].

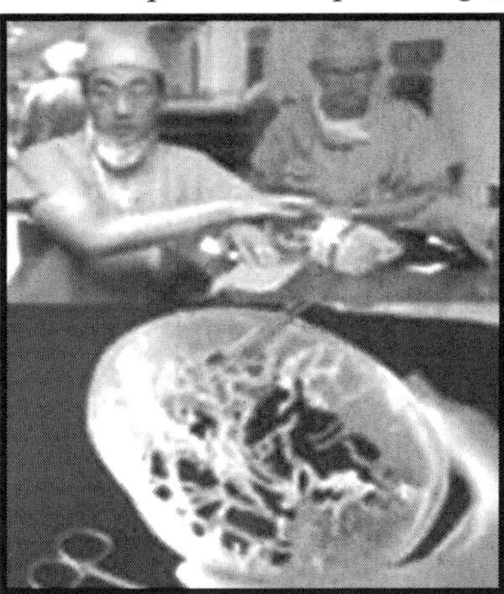

Fig. Biomodel use in presurgery planning

Fig. Left: CAD image- Right: Medical model

Customized orthoses and protheses implant/replacement

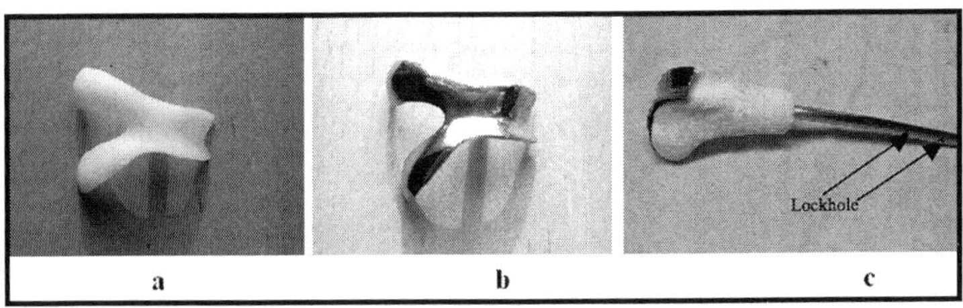

Fig. SL pattern of a hemi-knee joint (a) as a master for the titanium joint (b) which implanted in the femur bone (c)

Since every patient is unique, LM systems are used in fabrication of personalised implants for reconstructive and plastic surgery. Due to the inherent strength of LM technologies to fabricate complex geometry, it is very easy to manufacture custom implants. The model can be used as the custom implant itself or of the implant. Winder et al. [37] and D'Urso et al. [7] successfully used implants model fabricated using LM systems as a master model for reconstructive surgery of a skull defect. They claim there is a reduction on operating time and excellent outcome at 'reasonable' cost. The capability of LM to customise implants to quickly fit into a patient's unique size is a great advantage.

Hip sockets, knee joints and spinal implants could greatly benefit from this. He et al. [15] claims that the composite hemi-knee joint prosthesis (Fig. 4) reconstructed using LM are accurate to within a maximum tolerance of 0.206 mm. It fitted well and matched with the surrounding tissues, in particular to the lower tibial knee joint. Chang et al [4], Eggbeer et al [9], Kruth et al [21] and Bibb et al [2] had demonstrated the use of LM technologies in dental applications to be viable as it can improved the speed, quality and efficiency.

Scaffoldings and tissue engineering

With it ease of fabricating internal structure, LM technologies are ideal for generating implants with special geometrical characteristics, such as scaffolds for the restoration of tissues [18] as shown in Fig. 5. Scaffolds are porous supporting structures used as transplantation of tissue cells for the rapid and guided growth of new tissue in damaged or defective bones of the patient.

3D scaffold fabrication techniques for tissue engineering has been used for the last 30 years with less favourable success due to the lack of mechanical strength, no assurance of interconnection channels and uncontrolled pore size [18, 40]. Hence in order to for tissue reconstruction, scaffolds must have interconnected macro and micro-channels with high porosity and adequate pore size to facilitate cell seeding and diffusion of both cells and nutrients throughout the whole structure. These characteristics are required in producing scaffold for tissue engineering. LM has been studied by numerous researchers as a fabrication choice in constructing 3D scaffolds to guide the development tissue culture both in vivo and in vitro [13, 17, 33]. Armillota et al. [1], Hollister [17], Hutmacher et al [18], and Yang et al [40] have demonstrated and rationalised the use of LM technologies in scaffold design for tissue engineering to be viable, cost effective and practical.

Yang et al [24] state the advantages of using this technique in designing and fabricating scaffolds. Fig. 6 illustrates a scaffold tissue engineering proposed for reconstruction of a mandibular implant. LM systems like SLS, FDM, and 3DP have proved to be suitable for fabricating controlled porous structures for use in tissue engineering. LM technologies have contributed significantly to the field of scaffoldings and tissue engineering through the use of biomaterials [20].

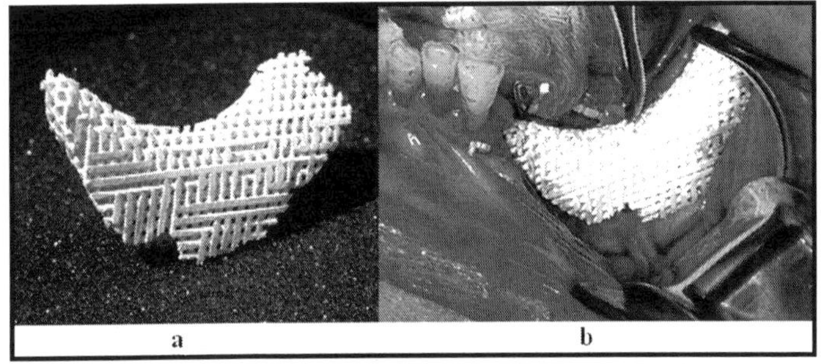

Fig. Layered manufactured scaffold tissue engineering (a) used in reconstruction of mandibular Implant (b)

Medical devices and Drug delivery systems

Another application of LM techniques is in fabricating medical devices and drug delivery systems. Skull defect and dental implant are restoration process that requires detailed planning and high accuracy in implant placement. Hence surgical guidance aids are required. Sarment et al. [27] and Di Giacomo et al. [6] investigated the use of SL surgical guides (Fig. 7) to accurately place dental implants and concluded that there is a significant improvement in implant placement.

Ruppin et al [29] claim that LM fabricated surgical guides are comparable to optical tracking system and in agreement other researches on accuracy in computer aided surgery for implant. Bibb et al [2] studied the use of LM to fabricated removable partial dental (RPD) framework for retaining artificial replacement teeth in the oral cavity. The patterns produced were deemed by a qualified and experienced dental technician to be a satisfactory fit and comparable with those produced by expert pattern technicians. In the study, the stiffer patterns produced by SL were easy to handle, were accurate, and produced satisfactory results. And Tay et al. [35] claim that an actual prosthetic socket fabricated using an FDM system (Fig. 8) provides an acceptable degree of comfort, and clinical trial confirmed the viability of fabricate prosthetic socket using FDM technology. Besides medical devices, LM methods are also use to produce drug-delivery system like oral tablet. Rowe et al [28], Leong et al [22] and Low et al [24]demonstrated the possibility of building oral tablet that controlled specific and precise drug delivery by using SL.

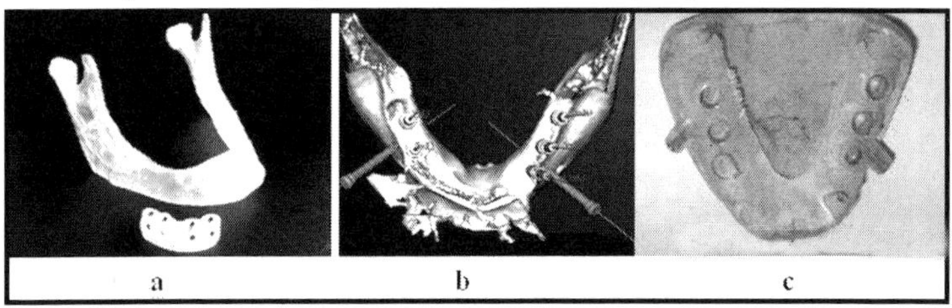

Fig. SLA surgical guide (a) and the constructed surgical guide in CAD (b) fabricated guide (c)

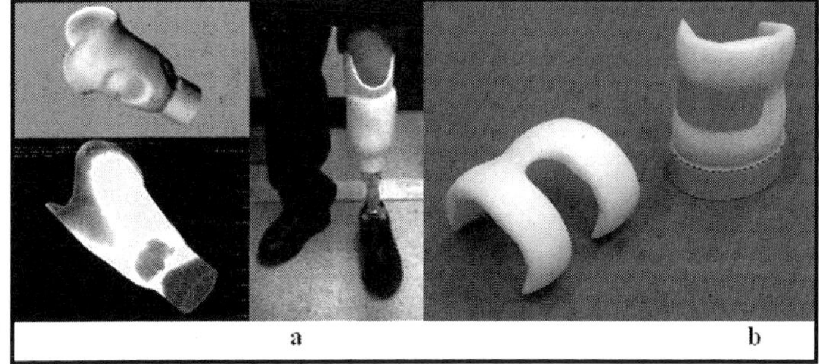

Fig. Below-the-knee Selective laser sintering prosthetic (a) [44] and FDM total knee replacement (b) [41]

COMPLETION OF LM AND CAD IN MEDICAL APPLICATIONS

The development of medical images into 3D models as a tool to help practitioners visualize 2D images has contributed to the development of a new methodology in fabricating medical parts. Medical imaging provides important data of various body structures for diagnostic reasons. These data can be used to obtain geometrical information of the body structures for three-dimensional modeling. CT and MRI images of various structures from conventional hospital scanners provided the input data for commercially available software packages. The image data are visualised, segmented and three dimensionally reconstructed. Solid models can then be generated for use in CAD systems. The generated models incorporating tissues of interest can be imported into a CAE environment for further CAD modelling and finite element analysis. That environment also serves as a platform for conversion to a readable format by rapid prototyping systems. LM systems are then used to produce the physical medical models.

The integration of Medical Imaging, CAD, FEA and LM has been presented as a realistic method for modeling and designing various body structures in medical applications. Hieu et al [38], Gopakumar [14] and Lohfeld et al [23] in the study of designing cranial and maxillofacial implant claim that there is a reduction time in implementing the integrated approach of Medical Imaging, CAD, FEA and LM for fabricating personalised medical implants. The common theme in Hieu et al [16], Gopakumar [14] and Lohfeld et al [23] methodology of this approach.

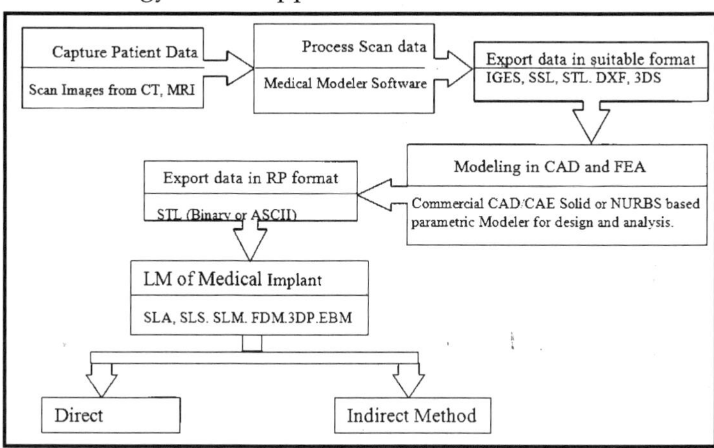

Fig. LM and CAD methodology adapted to design and manufacture of implants

However Starly et al [33] highlight a more comprehensive approach that generate CAD models from scan images. Integrated slice software in LM is used as an interface between the STL file generated from CAD modeled implants and the LM machine. It allows the user to specify the attributes for the LM to build the physical model.

IMPLANTS DESIGN FOR MANDIBULAR BONE BASED ON BIOCERAMICS AND NEW FIXATION METHODS

Bone is a living tissue; considered as a composite material, it comprises of trabecular (cancellous) bone and cortical (compact) bone. It has the capability of healing and remodelling. It will respond with adaptation in its structure to loading stress or injury such as fracture. However, bone is unlikely to remodel itself in major losses cause by trauma, cancer, congenital abnormalities or bone deficiency. Most of these types of major bone repairs are treated by grafting which uses the patient's own bone (autografts) or donor bone (allografts). The need of further surgery, risk of transmitted disease and limited material from donor site pose some limitations to the current practices [25, 34, 32]. Synthetic substitutes using metal, ceramic, polymer and composites are currently been use to overcome these limitations. Bioceramics are most frequently used in scaffold manufacturing and hard tissue implants within bones, joints and teeth. Bioceramics have the basic chemical composition akin to natural bone. Goodridge et al [13] and Dyson et al [8] have demonstrated the feasibility of using bioceramics for use as bone substitute. The design of the implant must take in to consideration biocompatibility, mechanical properties, cost effective manufacturability and process [19] as well as an accurate fit that requires minimal or no healthy bone removal. Furthermore, the bone variations in material and mechanical properties are dependence on location and function [12, 31]. Hence a promising method of manufacturing implants is through LM. This paper proposes a LM and CAD methodology to design customised implants of mandibular bone.

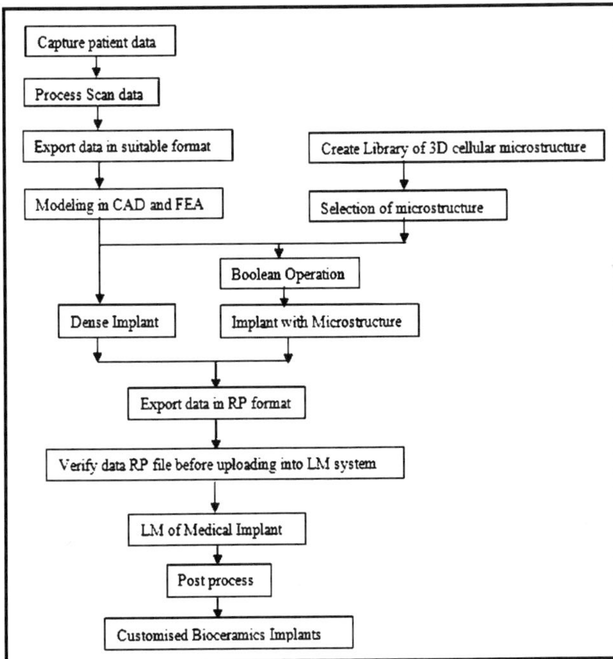

Fig. LM and CAD methodology for implants design in mandibular bone based on bioceramics and new fixation

Normally most of the fixation methods for mandibular reconstruction and fracture system consist of drill bits, plate bending forceps, plate holding forceps, plate cutters, cannulae, taps, countersinks, plate bending pliers, plate cutters, drill guides and screwdrivers to facilitate the placement of screws and modification of plates. The implant for reconstruction is secured in place by plate and screw. However, this paper proposed a design that integrates the fixation method into the implant. This is derived from the advantages of LM. A lower jaw bone model in STL format was imported to the CAD to design an implant. The lower jaw is non-defective and was obtained from a secondary source. A simulated defective section was created on the jaw by cutting a section of it in the CAD software.

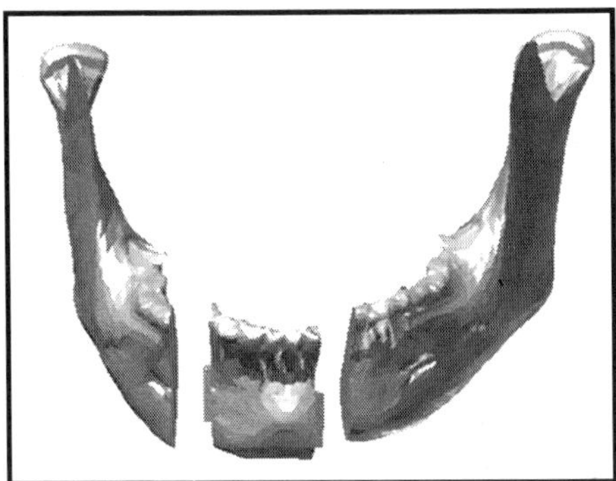

Fig. A section cut out for implant.

Several joints are proposed. Type (a), (b), and (c) joints rely on rapid bonding to secure the implant while type (d) uses conventional screws.

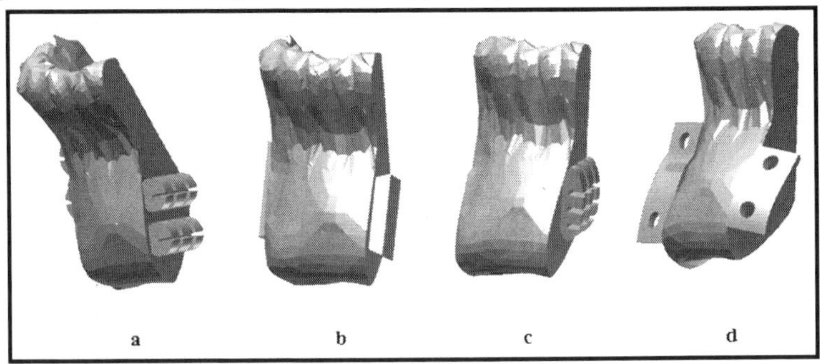

Fig. Different types of joint under consideration.

The design of the implants will need to incorporate interconnected controlled channels in order to promote bone growth. Bignon et al [3], Woesz et al [38] and Chu et al [5] indicated that there are significant correlation between micro- and macro-porosity as well as controlled channels with bone response and mechanical properties.

Conclusion and Future work

The LM and CAD approach for medical applications had proved to be viable and promises potential benefits as demonstrated by the numerous researches conducted. Precise customised bioceramic implants can reduce the removal of healthy bone, eliminate the need for bone grafting, and promote effective planning of implantation.

The future work to be undertaken should be related to finding suitable bioceramics' material and mechanical properties that adhere to the requirements of successful bone growth. In addition, further work will be conducted to propose new fixation methods and assembly. This will involved FEA to determine the strength of the propose fixation joint as well as of the implant itself.

4

Advanced Manufacturing Planning and Scheduling

PLANNING SYSTEMS

In this paragraph all the historic planning systems will be described briefly, starting with statistical inventory control (SIC). After the description of Material Requirements Planning (MRP I), Manufacturing Resources Planning (MRP II), Distribution Resources Planning (DRP) and Enterprise Resources Planning (ERP), this paragraph will end with a short description of APS.

STATISTICAL INVENTORY CONTROL

SIC is static in nature and operates solely on the basis of a predicted forecast. This method of inventory management employs a number of mathematical techniques to control inventories, based on historical turnover data. This method of inventory management is easy to computerise.

MATERIAL REQUIREMENTS PLANNING

The computerised data-processing techniques introduced in enterprises from 1950 made it possible to perform complex calculations and to process large amounts of data. In this period MRP I systems were developed. For the first time the factor 'time' made its entry into inventory management. MRP I systems operate on the basic of the existence of so-called dependent demand that can be calculated from a requirement for a product with an independent, predictable demand and the factor time in controlling inventories.

MRP I comprises a number of information-science techniques to plan material acquisition (the inflow of the necessary raw and auxiliary materials and semi-manufactures) and the production process on the basis of an established production plan for end products. A production plan is determined on the basis of market and turnover expectations. The composition of each product in terms of components (raw materials, auxiliary materials and semi-manufactures) is known and set out in a bill of material.

Given an established production program for a specific period, the planner uses MRP I to calculate which components are required in what quantities and at what point in time, by examining the throughput time or delivery time of the component (scheduling).

MANUFACTURING RESOURCES PLANNING

MRP II is an extension of MRP I, which assumes unlimited capacity. The extension to MRP II involved the calculation of the required capacity. On the basis of a required production program, MRP II calculates back from the delivery data to determine what capacity is required in what quantity and at what point in time in order to deliver the orders punctually. It is important to know at an early stage which capacity element in the process (machine, people, money, supplier, etc.) will constitute the bottleneck and when.

DISTRIBUTION RESOURCES PLANNING

A distribution network consists for the most part of several consecutive inventory points; for example the factory, a central distribution centre (DC) and national sales warehouses. In a distribution network, co-ordination of the various activities (sales forecast, orders, transport and inventories) is essential. The principles of MRP I/II (dependent demand and scheduling) are also used in inventory management in distribution networks: DRP.

DRP is an information system that supports co-ordination within the distribution network. The purpose of such a system is to record goods flows and it requires that information must be available on where stocks are held, which goods are in transit and what are the changes in inventories. DRP makes it possible to co-ordinate the decisions taken at various point in the distribution network.

Enterprise Resources Planning

ERP is defined as a software architecture that facilitates the flow of information between all functions within a company such as manufacturing, logistics, finance and human resources (Hicks, 1997). It is an enterprise-wide information system solution (Lieber, 1995). An enterprise-wide database, operating on a common platform, interacts with an integrated set of applications, consolidating all business operations in a single computing environment (Peoplesoft, 1997).

Ideally, the goal of an ERP system is to be able to have information entered into the computer system once and only once (Lieber, 1995). For example, a sales representative enters an order into the company's ERP system. When the factory begins assembling the order, shipping can check on the programs to date and estimate the expected transport date. The warehouse can check to see if the order can be filled from inventory and notify production of the number of products still needed. Once the order gets shipped, the information goes directly into the sales report for upper management. ERP provides a backbone for the enterprise. It allows a company to standardise its information systems. Depending on the applications, ERP can handle a range of tasks from keeping track of manufacturing levels to balancing the books in accounting. The result is an organisation that has streamlined the data flow between different parts of business (Lieber, 1995). In essence, ERP systems get the right information to the right people at the right time (Sheridan, 1995).

As a result of 'island automation' of individual parts of a company there are hardly, if any, links between those parts. However staff of one department need a better understanding of other departments' processes. ERP systems are helpful in this context. These systems take care of the entire administrative process of the various units within a company. A company can use an ERP package to drive all processes, such a financial management, sales forecasting, purchasing, inventory management, production control, logistics, project management, service and maintenance. Examples of ERP systems are Baan, Oracle, JD Edwards and SAP.

ADVANCED PLANNING AND SCHEDULING

"An APS system is a system that suits like an umbrella over the entire chain, thus enabling it to extract real-time information from that chain, with which to calculate a feasible schedule, resulting in a fast, reliable response to the customer. With the help of APS it is now possible to answer customer enquiries within seconds. This is just one of the possibilities of APS. The suppliers of APS can demonstrate impressive results: after implementation of APS, better throughput times, delivery times, inventory levels and utilisation rates result in improved operating results and a higher level of customer service." (Van Amstel et al., 1998).

There are two reasons why the interest and demand in APS systems arises at the moment. The first is the development of memory resident servers. Memory resident means that the entire planning engine, model and database are kept entirely in memory. This means very complex manufacturing and supply chain operation models can be stored in memory totally. This development provides a major advantage, because it eliminates disk access time and that gives serious time reduction in solving the planning problems. It allows very fast processing of large datasets, which makes simultaneous material and capacity problem solving possible (Bermudez, 1998).

The second reason is that companies are uniting their supply chains. Companies start to understand how the value chain works. Co-operating companies should manage their supply chains in one process. APS systems make it possible to co-ordinate these different supply chains in one system. System suppliers that successfully evolved to this level of planning and scheduling did so because they broke out of the traditional factory-only or distribution-only focus (Grackin, 1998).

APS is a new revolutionary step in enterprise and inter-enterprise planning. It is revolutionary, due to the technology and because APS utilises planning and scheduling techniques that consider a wide range of constraints to produce an optimised plan:

- Material availability
- Machine and labour capacity
- Customer service level requirements (due dates)
- Inventory safety stock levels
- Cost
- Distribution requirements
- Sequencing for set-up efficiency

PLANNING SYSTEMS VERSUS SUPPLY CHAIN INTEGRATION

In this paragraph the planning systems will be classified in a diagram, which is shown in below:

Table. Classification of planning systems in a environment/complexity diagram

Complexity / Environment	Functional	Integrated within	Integrated outside
static	SIC MRP DRP	ERP APS	APS
dynamic		APS	APS

The two axes of the diagram are:

1. *Environment:* The difference between static and dynamic is the level of predictability of the environment. In a static environment there is no need to reschedule or recalculate the plans that are made, because the environment is highly predictable. The organisation is familiar with the (number of) required products for the next period. Therefore it is enough to do the planning or scheduling at pre-defined times for a pre-set period. Instead, in a dynamic environment this predictability is very low. Due to this low predictability it is necessary to be able to reschedule plans very easily, and on a minute to minute basis.
2. *Complexity:* The complexity is divided in three layers of integration. The first layer is a "functional" organisation. In these kind of organisations the departments try to optimise their own department, without considering that it may not be optimal to the whole organisation.

The second layer is "integrated within" one organisation. In this layer a company is process driven and integrated. No outside information is gathered to optimise the planning. A separate organisation is an organisation with own profit/loss responsibility.

The third layer is "integrated outside" the organisation. When information of a production site with own responsibility for profit/loss is shared with the sales-organisation, these organisation is an "outside integrated" organisation. In the following subparagraphs the planning systems will be classified in the diagram with the axes environment and complexity.

SIC

This planning system will only function in a static environment in a "functional" organisation, because of the limited possibilities of this planning system. Some of these limitations of SIC are:

- Future requirements cannot always be predicted on the basis of historical data
- The specialist know-how that the planners have acquired are not used in the purely statistical approach to inventories

Due to these limited possibilities it is only possible to use SIC in a static environment. It is also not possible to use it for complex problems. Another disadvantage of SIC is that it results in the Forrester-effect. This effect is the result of the fact that different parts of the supply chain make independent decisions about inventories on the basis of its own stock calculation methods, which are static. These independent decisions result in higher and unbalanced stocks in the whole chain (Forrester, 1958).

MRP I/MRP II and DRP

These planning systems are now still operational in many organisations. In the functional organisation the planning is done separately for the various links in the chain. The planning is executed sequentially. The systems can only handle environments that are static and therefore also result in the Forrester-effect, because the various types of planning Master Production Schedule (MPS), MRP, and Capacity Resources Planning (CRP) affect each other due to the sequential process. The output of for example the MPS is the input for the MRP I/II run.

ERP

An ERP system can function very well in an environment which is still very static. An ERP system is ideal in companies that want to integrate their information flow within the organisation. In multi-site companies this can be viewed by the procedures. Each site (or profit/loss companies) has its own ERP system. It optimises the information flow for only that single site. An ERP system can be seen as a database which is surrounded by all sorts of applications. The database is the device that makes the integration in that company possible.

APS

An APS system can function in a number of environments and types of complexity. When companies start to integrate within their organisation an APS tool can be helpful, because the MPS-MRP-CRP planning process can take place simultaneously. An APS tool really benefits companies integrating with outside organisations. The customer and suppliers are involved in driving the organisation's logistical chain. Logistical planning and sales are merging in order to be able to respond rapidly to market requirements. The APS tool can be helpful in dynamic environments, because it has the advantage of being really fast in recalculating the plans whenever necessary. Another benefit of this system is that it facilitates the combination of information of multiple sites and that it calculates an optimal plan for a complete supply chain.

ANALYSIS OF THE PLANNING AND SCHEDULING FUNCTIONALITY

As mentioned in the previous chapter, this chapter discusses three basic (mathematical) functionality's of planning and scheduling in Advanced Planning and Scheduling systems (APS): concurrent planning (unconstrained planning), constrained planning and optimisation.

APS FUNCTIONALITY

The planning option (plan class) in APS specifies whether the plan should be based on constraints (materials, resources or both) or a financial optimisation.

- Unconstrained planning. In this option a traditional MRP calculation is generated based on assumed infinite material and resource availability. An exception message informs when materials and resource capacities have been exceeded.
- Constrained planning. In this option the generated plan respects the specified constraints. This option produces a feasible, but not necessarily optimal, plan as no plan optimisation objectives or criteria's are considered.
- Optimisation. In this option an optimised and executable plan is generated based on plan objectives and constraints. The optimisation is entirely based on cost and profit, which means that (soft) constraints could be overruled if this will reduce the total costs.

The three plan classes will be further described in the following paragraphs.

UNCONSTRAINED PLANNING

Unconstrained planning is a traditional MRP/CRP explosion of the master production schedule. MRP aligns supply quantities and due dates with demand quantities and time-phased net requirements are calculated for every part. Statements of material availability and resource capacity are used to generate exception messages to align supply due dates with customer due dates. Demand priorities are included during the planning run to determine the appropriate relationships between supply and demand. The replenishment plan is based on assumed infinite material and resource availability and exception messages are used to alert in case of materials or capacity shortage (Carol, 1999).

Example of unconstrained planning

The demand in each period is matched with planned production. In the third period, planned production exceeds capacity. MRP exception messages would indicate this condition. The planner would probably consider shifting some orders to a later production period to reduce the workload in period 3, and would perhaps rerun the MRP calculation to see whether the change causes an overload elsewhere in the production system. This way of planning is used in a traditional ERP-system. To create a feasible plan, the planner has to smooth out both materials and capacity demand, which is time-consuming.

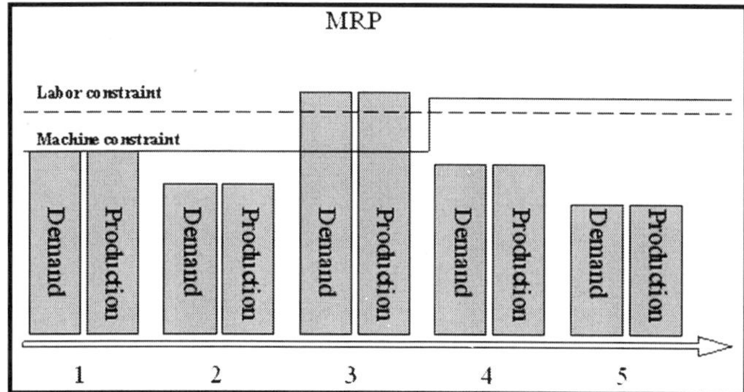

Fig. An example of unconstrained planning leading to overload of resources in period 3.

CONSTRAINT-BASED PLANNING

APS engines use constraints to help model the company specific manufacturing and distribution environment. Generally, constraints are a set of limitation, rules, and objectives that govern the physical and financial realm of possibilities for meeting the business plan (Bermudez, 1998).

- Limitations might include something as general as the availability of materials or machine capacity, or as detailed as the need for a minimum labour skill at a machine for a specific part.
- Rules might be as general as specifying that customer orders be considered ahead of forecast demand or as specific as the need to clean a machine after x number of production hours.
- Objectives are used to describe the company business plan and might include target safety stock levels, customer service levels, or sales revenue.

Rules are used as explicit decisions made by the planner and used when there are more options to choose in the plan generation. Rules are ranked by use of priorities. You can define and save rules based on the combination of criteria's such as dates, customer priorities, and item priorities (e.g. fill forecast with priority #1 ahead of sales order with priority #2.). Rules play an important role to gain benefit of advanced planning and scheduling.

Most APS products use some combination of limitations, rules and objectives as constraints (Lapide, 2000). The user might assign a target value to the constraint (when appropriate) as well as a weight to indicate the relative importance of this constraint. Some vendors have deployed slide bar controls, which work like the temperature control on a car's dashboard, to vary the weight assigned to each constraint. These slide bar controls also allow the constraint to be turned off. Other vendors would argue that slide bars are a gimmick because it is not practical to develop planning and scheduling algorithms that can consider infinitely variable constraints. Instead, this latter group of vendors controls the influence of the constraints in a number of different ways (Lapide, 2000):

- By turning the constraint on or off
- By changing the sequence in which the constraint is evaluated
- By assigning a specific weight or value
- By considering it hard or soft.

Generally, all advanced planning and scheduling vendors agree on the concept of soft and hard constraints. Hard constraints are usually physical limitations (can not be changed in the time horizon which is used in the planning process) such as limited machine capacity or material availability. Hard constraints are not overruled, whereas soft constraints are overruled, if necessary. As no plan optimisation objectives or criteria are considered this option produces a feasible, but not necessarily optimal plan. Soft constraints have no physical limitations and include business goals such as minimising set up cost, maintaining a target safety stock level or a required customer service level. When a product cannot be delivered on time the customer service requirements may be violated, but the product can still be delivered to the customer (Bermudez, 1998).

When the external domain is selected as a hard constraint, customer and supplier due dates are enforced while material and capacity availability are assumed infinite. When the internal domain is selected as a hard constraint, the capacity constraints are enforced to enable constraint-based planning, while demand due dates might be overruled. At the same time there is an option to determine whether the constrained plan enforces material, resources capacity or both. Resource capacity constraints are subdivided into operation resources, supplier resources, and transportation resources.

The opportunities to select different levels of constraint planning and different domains are an important functionality in APS. This makes it possible for companies to design their planning activities and structure according to their manufacturing condition and environment.

Most APS engines take a two or three pass approach to evaluating constraints (AMR, 1998). The first pass typically determines a feasible plan or schedule – one that tries to meet customer due date request without violating any hard constraints (this may be done in two phases in some products). In the second pass, the engine uses all of the constraints in an attempt to improve the plan or schedule. This second pass is generally referred to as optimisation (more on optimisation in the next section). Soft constraints may be used during this pass to solve for a better plan. Most APS products use an iterative, interactive approach that allows the planner to see the problems being encountered by the engine and to make decisions as to which constraints might be relaxed and by how much.

The distinction between hard and soft constraints is a matter of time horizon. Every constraint is soft, if the given time horizon is long enough. When the capacity is the problem and given time horizon allows capacity extension, the constraint is not hard, but soft.

Example of constraint-based planning

The example from figure (unconstrained planning) is illustrated using constraint/based planning.

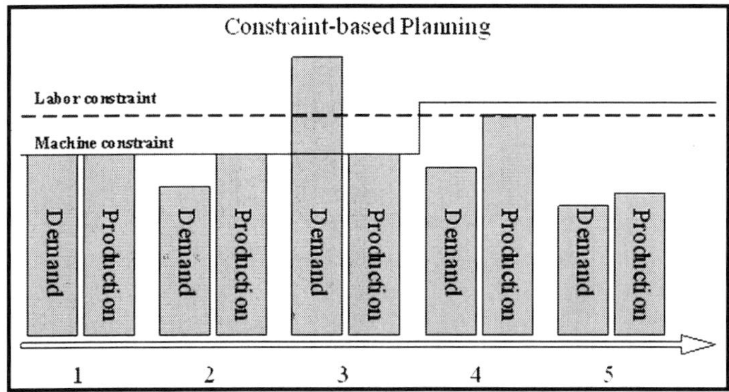

Fig. An example of constraint based planning, avoiding the resource overload form figure.

Supply exceeds in the second period, as inventory are accumulated for use during the third time period, in which demand substantially exceeds supply. Due to limited machine resources, demand can not be met and an order backlog occurs. Additional machine resources become available in the fourth time period, and production is now limited by labour availability. Demand is less than supply in this period, and some of the backlog is worked off, but not all of it. In the fifth time period, the remainder of the order backlog is worked off. Production is not constrained by either labour or machine resources.

In the above example, avoiding any backlog would be difficult, as it would require increasing machine capacity on a short-term basis. That is unlikely to be feasible. By working overtime in the fourth period, production could be increased to work off the backlog more quickly. This is a 'what-if' alternative that could be simulated by increasing the work hours for the labour resource.

OPTIMISATION

Today's market dynamics have made supply chains extremely complex and planning more difficult. Customer demand and competition have made supply chain planning and scheduling more challenging and complex. A number of major trends have contributed to this increasing complexity. These trends are contributing to an explosion in the number of entities that have to be planned for, driven by increases in the number of the following elements:

- Items
- Production and distribution facilities
- Functions
- Customers and suppliers.

For many years manufacturers have been moving toward improved use of technology to support complex, diverse planning processes (Lapide, 2000). Some are doing it largely to maintain control of their operations in order to meet customer demand. Having already achieved control, many manufacturers are using APS technology to increase the productivity of planning processes and to lower supply chain costs.

Generally, companies are looking for planning solutions that consider major supply chain constraints, which leads them to constraint-based optimisation. Supply chain planning optimisation techniques and solutions attempt to accomplish the following tasks:

- Determine a feasible plan that meets all demand needs and supply limitations
- Optimise the plan in relation to corporate goals such as low cost and profitability.

While a feasible, realistic plan is of paramount importance, and optimised plan is better. It is the need for realistic, optimised plans that is driving many manufacturers away from classic materials requirements planning (MRP)-based planning solutions, which do not consider supply constraints (especially material constraints) and frequently generate an unrealistic supply plan.

Consistent with this corporate trend toward greater need for supply chain planning technology, the APS market has increased dramatically between 1997 and 1998. Optimisation has been widely incorporated into APS suites. Examples include the following events:

- In 1997, Manugistics embedded various optimisation solution methods into its integrated supply chain-planning suite.
- In 1998, i2 Technologies extended its optimisation capabilities by

purchasing the CSC Operations Planning Group, which developed customised optimisation solutions for the consumer packaged goods (CPG) market. i2 Technologies also purchased Optimax Systems, a pioneer in the use of genetic algorithms to optimise the scheduling of assembly lines.

- SAP has developed the Advanced Planner Optimizer (APO) in 1999, which uses optimisation techniques.
- ILOG, INC, a supplier of supply chain optimisation software components to APS vendors, purchased CPLEX Optimization Inc., a supplier of linear and mixed integer programming tools, in 1998.
- Baan acquired Berclain in 1997, a production planning and scheduling vendor.

Despite the recent flurry created by the APS and ERP providers, it should be noted that supply chain planning optimisation technology solutions are not new. There has been a market for optimisation solutions for over 30 years. The market has slowly evolved from toolkit based products to a packaged application market. Early adopters of optimisation technology tended to be quantitative analysts, usually with degrees in operations research, who worked in the corporate world. Many worked in process industries such as Chemical, Paper and Steel. These early adopters used general-purpose optimisation tools (e.g., linear programming packages) purchased from software vendors to develop custom planning tools that typically ran in a batch mode.

As this market progressed, a few early supply chain-planning vendors started to sell general-purpose optimisation applications. These applications made it easier for corporate users to develop supply chain planning solutions on their own or working with the vendor's consultants.

Despite some early success in the use of optimisation, the market was relatively stagnant until recently. Advances in powerful computer technology have helped to accelerate the growth of the APS market. The technology has also allowed APS vendors to embed optimisation into their solutions more seamlessly and transparently. This has made it easier for users to model their planning environment, even those users not trained in optimisation techniques.

Today there are many popular APS solutions with embedded optimisation. The next paragraph describes the concept behind optimisation techniques and methods.

A supply chain optimisation problem

Generally, optimisation problems seek a solution where decisions need to be made in a constrained or limited resource environment. Most supply chain optimisation problems require matching demand and supply when one, the other, or both may be limited. By and large, the most important limited resource is the time needed to procure, make, or deliver something. Since the rate of procurement, production, distribution, and transportation resources is limited, demand cannot be instantaneously satisfied. It always takes some amount of time to satisfy demand, and this may not be quick enough unless supply is developed well in advance of demand. In addition to time, other resources, such as warehouse storage space or a truck's capacity, may be constrained in meeting demand. Optimised plans are generated based on plan objectives and constraints. The constraint-based rules are extended with some extra rules (titled decision variables and penalty factors). As the optimisation is based on cost and profit the constraints might be overruled if this reduces the total costs. For example, demand priority and supplier allocation ranks could be overruled to reach the best profit. If a rank 2 supplier results in lower cost than a rank 1 supplier, orders will be allocated to the rank 2 supplier. However all decisions can't be based on costs and profit. There could be many reasons that a supplier has a higher rank based on a total business point of view (e.g. better quality or better delivery performance). The total costs might be lower even though the costs of the part are higher, but this is not possible to model.

Decision Variables are within the planner's span of control.

- When and how much of a raw material to order from a supplier
- When to manufacture an order
- When and how much of the product to ship to a customer or distribution centre

Constraints are limitations placed upon the supply chain

- A supplier's capacity to produce raw materials or components
- A production line that can only run for a specified number of hours per day and a worker that must only work so much overtime
- A customer's or distribution centre's capacity to handle and process receipts

The constraints in an optimisation problem are either hard or soft. Most optimisation problem formulations designate cost penalties if a soft constraint is not met. The penalties allow constraints to be weighted by importance. For example, missing a customer due date is a more important concern than cluttering a warehouse aisle.

Objectives and implicit objectives

Objectives maximise, minimise, or satisfy something, such as the following:

- Maximising on/time delivery
- Maximising profits or margins
- Minimising supply chain costs or cycle times
- Maximising customer service
- Minimising lateness
- Maximising production throughput
- Satisfying all customer demand

Implicit objectives can be characterised as default or foundational objectives that the optimisation solver always attempts to honour. In addition to the objectives defined above, which can be selected/weighted or deselected by the planner, there is an implicit (hidden) objective that is taken into consideration no matter what the planner selects.

The implicit objective is maximised by minimising the penalty costs for:

- Late demand
- Supplier capacity violation
- Transport capacity violation
- Any unused supply
- Using alternate resources
- Unmet demand
- Resource capacity violation
- Safety stock violation
- Using alternate routings.

Implicit objectives are overridden if necessary when the primary objectives are specified. For example, to obtain the primary objective: on-time delivery, it could become necessary to substitute resources, bills, routings, or items. Other substitutes and alternates may also be recommended for cost saving reasons.

The optimised plan suggests which items to produce, how many to order, and the best time to order them. It also suggests the best source for the products, the best bills, routings, and resources to use, the best transportation methods, and the best level of safety stock inventory to maintain, all in relation to cost and profit.

The optimisation satisfies weighted objectives and takes into consideration the penalty factors related to these decision variables. The following penalty cost factors are used explicit in relation to decision variables:

- Late demand
- Exceeding resource capacity
- Exceeding material capacity
- Exceeding transportation resource capacity

The user enters percentages to indicate how important it is for him that those outcomes do not occur in his plan. The optimisation process drives penalties out of the solution, tending to drive the most costly penalty factors out first. A high degree of accuracy in setting penalty factors is not as important as the relationship between penalty factors.

When the system make decisions to avoid late demand, it will place higher priority on keeping large sales on time.

When the penalty for late demand is higher than the penalty for exceeding resource capacity (factor times work order resource cost), the solution will tend to plan overtime work in order to avoid late delivery. In general, all penalty factors work this way.

Objective weights in general do not show the precise relative importance of each objective in planning decisions. The percentage of the objective value occupied by a particular objective depends also on the dollar magnitude of the objective, and it is the product of the weight and the dollar magnitude of the objective which reflects the relative importance of each objective in planning decisions.

It is important to realise that the multiple objectives must have the same order of magnitude. At the same time, knowledge about both the production condition/structure and cost structure is vital as optimisation is based on relative parameters.

Example of optimisation

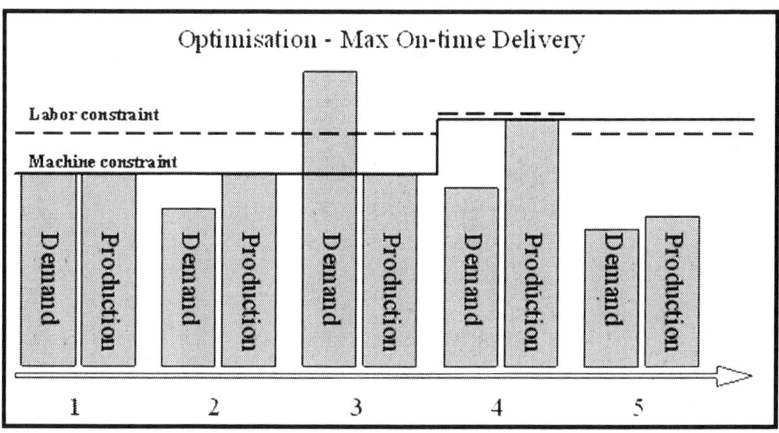

Fig. An example of an optimised plan in relation to on-time delivery.

In the first three periods of this example, there is no difference between the optimised plan and the constrained-based plan. A backlog occurs in the third period because the hard machine constraint makes it impossible to meet the peak demand.

However, production in the fourth time period has been increased compared to the CBP example. Recall that in the CBP example, some of the period 3 demand was backordered and not met until period 5. In the optimisation the cost of labour overtime in the fourth period is balanced against the cost of carrying the backorder into period 5. If the backorder quantity is large, and if the customer is likely to accept a two period delay, and if the cost of overtime is relatively low, then optimisation would suggest3.

Optimisation framework

The structure of the supply chain need to be represented. This is typically done using a network model which graphically visualises a supply chain and is used to depict the parts of a supply chain being considered in the planning process.

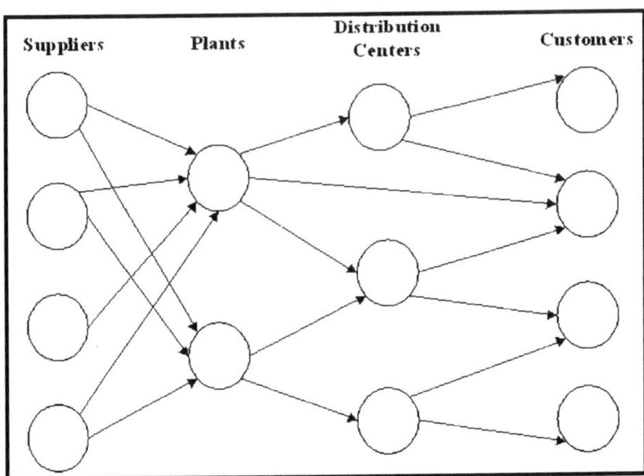

Fig. Network representation of a Supply Chain

Figure represents a manufacturer's supply chain (Kok, 2001; revised). Usually referred to as a network representation, the nodes represent facilities that add value to the supply chain. Nodes occur from the sources of raw materials and intermediate products to the consumers of the finished products. The arcs or links connecting the nodes represent transportation lanes for materials, semi-finished, and finished products.

Optimisation solvers

While it is safe to assume that every plant manager would like to have the optimal production schedule, there is plenty of controversy over exactly how a plan or schedule is optimised. To many manufacturers an optimised schedule is one that meets all customer due dates. To APS developers, optimisation is a systematic approach to improving the plan or schedule based on the constraints of the business. This differs from simply meeting due dates while considering soft constraints such as minimising inventory or maximising revenue. While APS vendors would not disagree on this definition, the techniques or algorithms used to solve for an optimised plan or schedule vary widely.

Some vendors attempt to achieve optimisation by applying a single algorithm to a wide range of problems, while others maintain a library of algorithms or "solvers" which can be used in a trial fit approach. As described in chapter 4 there are different techniques that can be used for optimisation:

- Linear programming – A complex algorithm that attempts to express the problem as a set of mathematical equations. This algorithm is sometimes used to establish a baseline plan. It is very popular for site selection or sourcing problems.
- Genetic Algorithms – A brute force method based on the concepts of genetics. Under this theory, it is believed that a species improves itself over time through the genetic combination of superior gene

pools – essentially, survival of the fittest. In genetic optimisation techniques, the best group of plans or schedules is selected from each iteration as the starting point for another optimisation pass.

- Theory of Constraints – A systematic method that attempts to move material quickly and smoothly through the production process in concert with market demand. Using three simple global measures, throughput, inventory, and operating expense, the production process is refined to achieve goals of the organisation within the market and production constraints.
- Heuristics – Another technique made feasible by the power of today's computer. This is essentially a trial and error approach that may look ahead or look backward to improve the plan or schedule.

With the possible exception of linear programming, most optimisation algorithms compare each new plan or schedule against an old one. In making this comparison, the algorithms must evaluate the various trade-offs between inventory, machine-utilisation and delivery performance in conjunction with other constraints. In order to make this comparison, each plan and schedule must be scored. Scoring may be based on costs, weighting factors, or units. Some vendors go as far as allowing the use of activity-based costs, while others assign relative costs or penalties. Still others use scorecards that list constraint violations, such as material shortages, inventory stock-outs or overdue orders. These scorecards allow the planner to visually assess the impact of changes to the plan or the constraints.

Optimisation techniques often compare thousands, if not hundreds of thousands of schedules to find the best one. Generally, most techniques reach a point of diminishing returns, where the potential incremental improvement in the plan is minuscule, and the time required to find each improvement grows exponentially. Some vendors graphically display the progress of the optimiser versus time and allow it to be stopped manually (Lapide, 2000). Stopping the optimisation process is not without risks. Some planning and scheduling environments are subject to a phenomenon called local optimisation. It is possible that the optimiser is stuck at a point where changes in either direction appear to produce inferior schedules. This effect can be caused by the batching rules for certain types of production equipment, among other thing.

Generally, mathematical programming methods are used in solvers for strategic and higher levels of tactical planning. These methods generally work only for solving linear- and some integer- based models, commonly used in strategic levels of planning. Tactical and operational models are usually not linear and are much too complex to solve using mathematical programming methods. For this reason, heuristic methods are generally used in tactical and operational planning level solvers. Genetic algorithms are used primarily in operational planning to consider a large number of possible solutions. The

Theory of Constraints, a heuristic method based on work by Eli Goldratt, is another solver commonly used in operational planning. Vendors that use solvers based on the Theory of Constraints are: i2 Technologies, STG and Thru-Put Technologies.

While not a formal optimisation technique, exhaustive enumeration is predicated on using the computer to find a solution by looking at all possible alternative plans. This method proves useful in simple supply chain situations. Otherwise, this method is computationally intensive and slow to generate a solution.

Distinction Software uses this optimisation method for its manufacturing planning solutions. Since the company focuses on mid-tier and smaller manufacturers, the exhaustive enumeration approach is feasible.

A standard LP-model for optimisation

Currently commercial SCP software assumes a rolling schedule concept, where each planning cycle a mathematical program is solved, either to optimality or some heuristics are applied. For uncapacitated SCP problems without lot sizing restrictions it is rather straightforward to formulate an LP model that fits in this rolling schedule context. In this paragraph we formulate a standard linear programming model in a rolling schedule context as applied to supply chain control problems.

Let us consider a supply chain where at the control level we deal with N items. For each item we define (Kok, 2001):

- LI THROUGHPUT TIME BETWEEN TIME OF RELEASE OF AN ORDER FOR ITEM I AND TIME AT WHICH THE ordered items are available for usage in other items and/or delivery to customers
- aij number of items i required to produce one item j
- DI(T) EXOGENOUS DEMAND FOR ITEM I IN PERIOD T, I.E. DEMAND IN PERIOD T FOR ITEM I, THAT IS NOT derived from demand for items in Ei \ {i}
- P set of products with exogenous demand, i.e. {i | ?t ? 1, Di(t) > 0}
- E Set of end products, i.e. {i | ?j, aij = 0}
- I Set of intermediate items, i.e. {i | ?j, aij > 0}
- ri(t) quantity of item i released at the start of period t, t?0, ?i
- Ji(t) net inventory of item i at the start of period t immediately before quantity released at the start of period t-Li is available, t?0. ?i
- Ii(t) physical inventory of item i at the start of period t immediately before quantity released at the start of period t-Li is available, t?0. ?i
- Bi(t) backlog of item i at the start of period t immediately before quantity released at the start of period t-Li is available, t?0. ?i

We assume that the incidence graph (aij) is acyclic and Li is constant. Then ri(t) must satisfy the following equations,

$$J_i(t+1) = J_i(t) - \sum_{j=1}^{N} a_{ij} r_{ij}(t) - D_i(t) + r_i(t - L_i), \forall i, t = 0,1,2,...,T \quad (1)$$

which is the inventory balance equation for general assembly networks. Furthermore ri(t) must satisfy the following inequalities,

$$\sum_{j=1}^{N} a_{ij} r_j(t) \leq \max(0, I_i(t) - B_i(t)) + r_i(t - L_i), \forall i, t = 0,1,2,...,T \quad (2)$$

$$r_i(t) \geq 0, \forall i, t = 0,1,2,...,T$$

It can be shown that is equivalent with,

$$B_i(t+1) - B_i(t) \leq D_i(t), \forall i, t = 0,1,2,...,T-1,$$

which states that the backlog from the start of a particular period to the start of the next period cannot increase more than the exogenous demand during this period.

In order to compare different supply chain planning concepts we define a cost structure and a performance criterion. We define C(t) as the cost incurred at the start of period t, $t \geq 0$,

$$C(t) = \sum_{i=1}^{N} h_i I_i(t),$$

where

h_i value of item i $\forall$ i

C(t) is not really a cost function but represent the total supply chain inventory capital investment at the start of period t. We are interested in the long-run average value of C(t),

$$\overline{C} = \lim_{t \to \infty} \frac{1}{t} \sum_{s=1}^{t} C(s).$$

The long-run average supply chain inventory holding cost can be derived from multiplying $\overline{C}$ by the interest rate. As performance criterion we choose $P_{1,i}, \forall i \in P$, defined as,

$$P_{1,i}$$

$$\lim_{t \to \infty} P\{I_i(t) > 0\}, \forall i \in P$$

For each supply chain-planning concept P we want to solve the following problem

$$\min \overline{C}(P)$$
$$P_{1,i}(P) \geq P_{1,i}^*, i = 1,2,...,N$$

We introduce the concept of safety stock in order to cope with short-term exogenous demand uncertainty, v_i safety stock parameter of item i,

$$i \in P$$

The safety stock parameters are used to control the end-item service levels.

Another issue to be dealt with is the mutual dependence of release decisions for different items. We can derive that a planning horizon T should be at least equal to the maximum cumulative planned lead-time as defined by the product structure.

In order to derive the systemwide order release decisions at the start of period t we need to forecast exogenous demand until period t+T-1. The solution to the MP problem not only provides us with the immediate order release decisions, but in addition provides us with planned order release decisions.

Therefore we define the following variables,

$\hat{D}_i(t,t+s)$ forecast of exogenous demand for item i in period t+s as decided on at the start of period t, $t \geq 0$, $S \geq 0$, $\forall\, i\, \hat{I}_i(t,t+s)$ forecast of physical inventory of item i at the start of period t+s as determined at the start of period t, $t \geq 0$, $s \geq 0$, $\forall\, i\, \hat{B}_i(t,t+s)$ forecast of backlog of item i at the start of period t+s as determined at the start of period t, $t \geq 0$, $s \geq 0$, $\forall\, i\, \hat{r}_j(t,t+s)$ forecast of quantity of item i released at the start of period t+s as determined at the start of period t, $t \geq 0$, $s \geq 0$, $\forall\, i$

Now we can formulate a Linear Programming model that can be solved by standard algorithms, such as the simplex method:

$$\min \sum_{i \in E} \left(\sum_{s=1}^{T} h_i Z_i^+(t,t+s) + \theta_1 h_i Z_i^-(t,t+s) + \theta_2 h_i \hat{B}_i(t,t+s) \right) + \sum_{i \in I} \sum_{s=1}^{T} h_i \hat{I}_i(t,t+s)$$

such that,

$$\hat{I}_i(t,t+s+1) - \hat{B}_i(t,t+s+1) = \hat{I}_i(t,t+s) - \hat{B}_i(t,t+s)$$

$$-\sum_{j=1}^{N} a_{ij} \hat{r}_j(t,t+s) - \hat{D}_i(t,t+s), s = 0,\ldots,T-1 \; + \hat{r}_i(t,t+s-L_i), \forall i$$

$$\hat{I}_i(t,t+s+1) - \hat{B}_i(t,t+s+1) = v_i + Z_i^+(t,t+s) - Z_i^-(t,t+s), \forall i, s = 0,\ldots,T-1$$

$$\hat{B}_i(t,t+s+1) - \hat{B}_i(t,t+s) \leq \hat{D}_i(t,t+s), \forall i, s = 0,\ldots T-1$$

We assume non-negativity of all decision variables involved in the LP-model above.

OPTIMISATION USAGE GUIDELINES

Though there are no hard-and-fast rules for manufacturers deciding whether to purchase optimisation technology, there are some guidelines (Lapide, 2000):

- Optimisation is generally beneficial in complex manufacturing environments where many interrelated decisions need to be made. These include environments with many resource constraints and large numbers of products, plants, suppliers, and distribution centres. Planners in these environments need computer support to make optimised decisions. In contrast, planners may not need optimisation support in simple, mature environments, where methods based on experience may already yield nearly optimal decisions.
- In strategic and higher-level tactical planning, the pursuit of optimised solutions is typically more important than it is for low-level tactical and operational planning. In the former, the feasible set for decisions is much larger, meaning there are more opportunities to make poor decisions. Also, these decisions have greater revenue and cost implications.
- The answer to "Where is the most pain in my supply chain?" will be important in deciding what portion to optimise. In supply-constrained industries that experience material shortages, optimising the use of these materials in the manufacturing process is important. In make-to-order environments, especially in discrete manufacturing, optimised production schedules are crucial. For distribution-intensive environments, planning must focus on optimising manufacturing and distribution operations simultaneously.
- Optimisation is more useful in mature, relatively non-volatile manufacturing industries where product demand and manufacturing processes are more predictable. In these planning environments, realistic models can be constructed to support all levels of planning. In volatile manufacturing environments, optimisation will be less useful for strategic and tactical planning. In these environments, planning focuses on supply chain readiness and responsiveness rather than on operational efficiency. Optimisation will be more useful for operational planning, when the level of uncertainty is substantially reduced (e.g., when many customer orders are already placed).

Uncertainty

The issue of uncertainty plays a role in planning in several ways. In production planning uncertainty exists in the order acceptation phase with respect to future orders, workload of main activities of accepted orders, release dates for materials or even regarding resources fitting the still unknown detailed design requirements of accepted orders. Non-regular capacity and order acceptance rules and due date setting can be used to cope with this uncertainty. In production scheduling the requested resource types and the capacities are known but workload and release dates still contain uncertainty. This will affect the on-time delivery service levels. Analogous sources of uncertainty are found in multi-level inventory management in supply chains due to upstream stock availability and in transportation when transportation times are stochastic due to congestion. For special problems we can derive optimal policies under uncertainty e.g. with regard to processing times or demand. Interestingly, such policies often differ from the optimal policies under certainty in a rolling schedule context.

The algorithms used in the described (traditional) planning systems and the most APS-systems use deterministic models and data. The deterministic planning's algorithms react fast at change, but assume flexibility and reserved capacity. In these deterministic models uncertain, variable, incomplete or even incorrect data is presented by the expected or worst-case value. Then sensitivity-analyses is applied afterwards. This is a reactive approach, because herewith only the impact of fluctuations in the data of the solution are studied. In practice this leads to nervous planning, that anticipates quasi real-time on changes. Many business-sectors can't cope with those changes because of technological and economical reasons, at least not without increasing costs.

To find solutions that are less sensitive to uncertainties of the parameters, a pro-active approach is needed. That means that uncertainties should be included in the model and that the algorithms should strive for specific reduction of the variability. This new approach is named 'Robust Planning' (Van Landeghem, 2000). These new mathematical models and optimisation-algorithms, explicit reckon with the variability and uncertainty of the relevant parameters in the supply chain and they generate more predictable and stable planning's.

Of the three planning levels in supply chains, the tactical level is the most suitable to deal with the causes of uncertainty. At the operational level there is to little time to react to fluctuations of uncertain parameters, and at the strategic level many phenomenon's are too variable to base a long-term decision on.

THE INTEGRATION OF THE SUPPLY CHAIN

"Like the medieval lords who built moats and walls around their castles many organisations have constructed artificial boundaries between themselves and the outside world. While these boundaries do not consist of water and bricks, they are just as difficult to surmount. More importantly, just as social evolution made castle walls obsolete, the new success factors of speed, flexibility, integration, and innovation are making boundaries between organisations less relevant. In fact, hiding behind such boundaries today can be more dangerous than venturing outside." (Ashkenas et al., 1995)

SUPPLY CHAIN

World class companies are now accelerating their efforts to align processes and information flows through their entire value-adding network to meet the rising expectations of a demanding marketplace (Quinn, 1993).

Some of the drivers for change, that forces companies to overhaul their logistical structure are (Holmes, 1995):

- *Increased regional and global competition:* The most potent force driving companies to overhaul their supply chains is increased crossborder competition, regional and global. For many companies the competitive arena has become worldwide, rather than national or regional.
- *The role of the single market in Europe:* Europe's single market has intensified competition by tearing down the last protective barriers. At the same time the single market is an important factor which enables supply chain integration across borders. The dismantling of frontier controls has led to the speed-up of road transport, which facilitates the switch from national to multi-country distribution centres.
- *Shorter product life cycles:* The trend towards shrinking product life cycles force a change in logistic management as it augments the risk of being stuck with obsolete inventory.
- *Changes in the market place:* National and crossborder mergers and acquisitions in recent years have led to greater concentration of purchasing power in most sectors of industry. In the wholesale and retail distribution the growth of powerful chains is squeezing out the independents.
- *Pressure from smarter customers:* Major retailers and industrial end-users are becoming more sophisticated and more demanding. They are reducing their supplier base and are working more closely with the remaining suppliers.
- *Service as a differentiator:* Products are more and more becoming commodities, forcing suppliers to search for new ways to differentiate themselves. Competitive edge will come from service differentiation.

The ability of an organisation to distinguish itself is coming to lie increasingly in the area of customer service. This places heavy pressure on the logistical chain. Delivering goods to customers in the most economic way while providing first-class service and quality is the logistics strategy. This requires more and more integration of the supply chain, in which all parts of the supply chain are linked to each other.

Suppliers and customers cannot be managed in isolation anymore, with each entity treated as an independent entity. More and more, there is a transformation in which suppliers and customers are inextricably linked throughout the entire sequence of events which brings raw material from its source of supply, through different value-adding activities to the ultimate customer. Success is no longer measured by a single transaction; competition is now evaluated as a network of co-operating companies competing with other firms along the entire supply chain.

Analytically, a supply chain is simply a network of material processing cells with the following characteristics: supply, transformation and demand (Davis, 1993). An example of a supply chain is shown in below

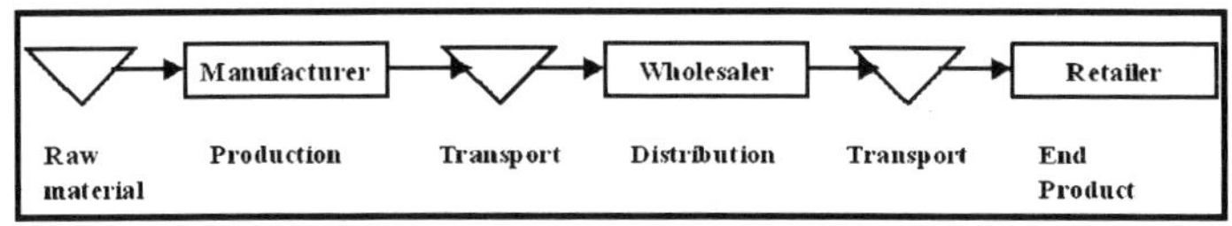

Fig. An example of a supply chain

SUPPLY CHAIN MANAGEMENT

Supply chain management (SCM) is defined as a process for designing, developing, optimising and managing the internal and external components of the supply system, including material supply, the transformation of materials and distribution of finished products or services to customers, that is consistent with overall objectives and strategies (Spekman et al., 1998).

The essence of SCM is to develop a sustainable competitive advantage by reducing investments without sacrificing customer satisfaction (Lee & Billington, 1992). Since each level of the supply chain focuses on a compatible set of objectives, redundant activities and duplicated efforts can be eliminated (Spekman et al., 1998).

In addition, supply chain partners share information that facilitates their ability to jointly meet end-users′ needs (Spekman et al., 1998). IT is an enabler and a key to the development of an integrated supply chain. However, this information must be shared by the partners. Research (Spekman et al., 1998) seems to suggest that there is a reluctance to share key information among partners. Many of these fears subside if partners share similar values and a common vision. Such information sharing heightens the alignment between partners such that effective supply chains share learning's among partners rather than worry about knowledge expropriation. The goal is to orchestrate this alignment and to ensure that the supply chain is better than the sum of its parts. Adopting the concepts and tenets of SCM requires a new mindset. SCM requires to look at the complete set of linkages that tie suppliers and customers throughout the supply chain.

SUPPLY CHAIN INTEGRATION

All companies function as links in chains of entities that produce and distribute products. Many companies have viewed their participation in the supply chain from an independent perspective, and have focused on the maximisation of its own profitability. This traditional view leads to the following types of boundaries in the supply chain, which reduce competitiveness by reducing speed, flexibility, integration and innovation (Ashkenas et al., 1995):

- *Strategies and plans are developed independently:* Each separate organisation has its own market targets, production plan, and schedule. The other parts in the supply chain are not consulted, which results in an unsynchronised supply chain.
- *Information sharing and joint problem solving are limited:* Organisations withhold information about cost price, profit margins, and problems from other parties in the supply chain. The tendency is to solve these problems alone, often resulting in suboptimal solutions or delayed product delivery.
- *Resources are utilised inefficiently:* In the different parts of the supply chain a lot of resources, expertise and knowledge is held separate from the other parts of the supply chain. All these separate parts use their own resources only for themselves, without the possibility of any other part to use these resources when they are temporarily superfluous.
- *Accounting, measurement, and reward systems are separate and unsynchronised:* Each part of the supply chain has its own accounting, measurement and reward system. Some parts emphasise on quality and others emphasise on sales volume.
- *Salesforce pushes products on salespeople's terms:* Salespeople focus on pushing products to the customers, while each part of the supply chain aims to maximise its own profitability. These salespeople do not listen to the requirements set by the customer which results in dissatisfied customers.

Successful companies will be those that take a systematic, boundaryless view of their participation in the supply chain. They must acquire an entirely new mindset, abandoning the legalistic view of organisations as independent entities linked only by market forces and learning to see themselves as part of an integrated system. By making specific external boundaries more permeable, organisation can dramatically increase speed, flexibility, integration and innovation (Ashkenas, 1995).

In the traditional view each organisation aims to maximise its own profit, while in the new model each organisation aims to maximise total supply chain success. The company in the new model will loosen its external boundaries and will follow a new model (Ashkenas et al., 1995):

- *Business and operational planning are co-ordinated:* In the successful supply chain, all members collaborate in both strategic and operational business planning. The goal is not only better product development and production planning, but also common or coordinated administrative and operational procedures such as billing, customer service, purchasing, shipping and inventory.

- *Information is widely shared and problems are solved jointly:* As members of a system, participants in a boundaryless supply chain share information more freely than before. A production problem in one part of the chain is everyone's concern, and the best resources throughout the system are applied.
- *Resources are shared:* A systematic view of the supply chain allows companies to deploy resources and expertise more efficiently throughout the chain.
- *Accounting, measurement and reward systems are consistent:* A key requirement for a boundaryless supplier-customer relationship is a common score-keeping and incentive system so that everyone in the supply chain works off the same numbers, speaks the same language, and aims towards the same set of goals. Successful supply chains have jointly accepted methods to determine costs, margins and investments. Agreed-upon performance goals for each organisation unit are derived from those methods. A matching reward system motivates employees to achieve the system-wide objectives.
- *Selling is a consultative process:* In the boundaryless world, successful companies engineer a significant shift in the role of their salespeople. Instead of pushing products, salespeople increasingly consult the customer, helping customers crystallise supply chain requirements and find optimal ways to meet those requirements and best utilise purchased products. In short, salespeople create a pull for a product.

Table. Overview traditional and new model view

Traditional View	New Model View
Strategies and plans are developed independently	Business and operational planning are co-ordinated
Information sharing and joint problem solving are limited	Information is widely shared and problems are solved jointly
Resources are utilised inefficiently	Resources are shared
Accounting, measurement, and reward systems are separate and unsynchronised	Accounting, measurement, and reward systems are consistent
Salesforce pushes product on salespeople's terms	Selling is a consultative process

Four forms of supply chain integration can be distinguished (Boorsma & Van Noord, 1992):

- *Physical integration:* Physical integration can be defined as those activities that focus on the improvement of efficiency of the primary process, by which the logistical costs of this process decrease, between minimal two entities in the supply chain. An example of physical integration is the use of standardised transportation devices.

- *Information integration:* A second form of supply chain integration are activities to attune the flow of information. As with physical integration, the primitive form of the logistical process and the management system do not change. An example of information integration is to forward shipping information from shipper to transporter.
- *Management control integration:* Management information, out of other entities in the supply chain, is used in a systematic way to integrate several parts of the supply chain. The goal is not only to generate cost benefits, but also to realise a better customer service level. By connecting the management information between entities in the supply chain, the total supply chain can respond quicker and more effective to the market requirements. An example of this integration is a supplier who receives information from its customer about the inventory level of a specific product.
- *Organisational integration:* Parts of the management activities come to lie at another entity in the supply chain. This concerns more than just the outsourcing of operational activities. It concerns the assignment of logistical planning tasks. An example of organisational integration is a company which partly takes care of the production planning.

ADVANCED PLANNING AND SCHEDULING

An overview of APS is given. First the different APS solutions which can be distinguished. The difference between enterprise and plant-centric systems are explained, the difference between advanced planning and advanced scheduling.

APS SOLUTIONS

APS can be viewed as an umbrella technology which uses a number of features which are described in paragraph 4.3. The scope of APS is not limited to factory planning and scheduling. It includes a full spectrum of solutions, both enterprise and inter-enterprise planning and scheduling systems. Differences are not only the time horizon, but also the level of the planning horizon, such as strategic, tactical or operational planning is considered. Based on Advanced Manufacturing Research (Bermudez, 1998), the following solutions can be distinguished:

Strategic and long-term planning

This solution addresses issues like:

- Which products should be made?

- What markets should the company pursue?
- How should conflicting goals be resolved?
- How should assets be deployed for the best ROI?

Supply chain network design

This solution optimises the use of resources across the current network of suppliers, customers, manufacturing locations and DCs. What-if analyses can be performed to test the impact of decisions to open new or move existing facilities on profit and customer-service level. It can also be a helpful tool to determine where a new facility should be located to fulfil customer demand in the most optimal way. These supply chain network design tools are mostly applied to find the balance between holding more stock at a specific location or making more transportation costs.

Demand planning and forecasting

Both statistical and time-series mathematics are used in this solution to calculate a forecast based on sales history. A demand forecast is unconstrained because it considers only what customers want and not what can be produced. Based on the information from the forecast, it is possible to create more demand through promotions in periods where the demand is less than maximum production.

Sales and operations planning

This is the process which converts the demand forecast into a feasible operating plan which can be used by both sales and production. This process can include the use of a manufacturing planning and/or a supply chain network optimising solution to determine if the forecast demand can be met.

Inventory planning

This solution determines the optimal levels and locations of finished goods inventory to achieve the desired customer service levels. In essence, this means that it calculates the optimal level of safety stock at each location.

Supply chain planning (SCP)

SCP compares the forecast with actual demand to develop a multi-plant constrained master schedule, based on aggregate-level resources and critical materials. The schedule spans multiple manufacturing and distribution sites to synchronise and optimise the use of manufacturing, distribution and transportation resources.

Manufacturing Planning

Develops a constrained master schedule for a single plant based on material availability, plant capacity and other business objectives. The manufacturing planning cycle is often only executed for critical materials, but that does depend on the complexity of the bill of material. Also the desired replanning time is a factor that one must take into account when deciding which level of detail is used. For example, with a simple bill of material a complete MRP I/II explosion can be executed in a few minutes.

Distribution Planning

Based on actual transportation costs and material allocation requirements a feasible plan on the distribution of finished goods inventory to different stocking point or customers, is generated to meet forecast and actual demand. With this solution it is possible to support Vendor Managed Inventory.

Transportation Planning

A solution which uses current freight rates to minimise shipping costs. Also optimisation of outbound and inbound material flow is used to minimise transportation costs or to maximise the utilisation of the truck fleet. Another possibility is to consolidate shipments into full truckloads and to optimise transportation routes by sequencing the delivery/pickup locations.

Production Scheduling

Based on detailed product attributes, work centre capabilities and material flow, a schedule is determined that optimises the sequence and routings of production orders on the shop floor.

Shipment Scheduling

This solution determines a feasible shipment schedule to meet customer due dates. It determines the optimal method and time to ship the order taking customer due dates into account.

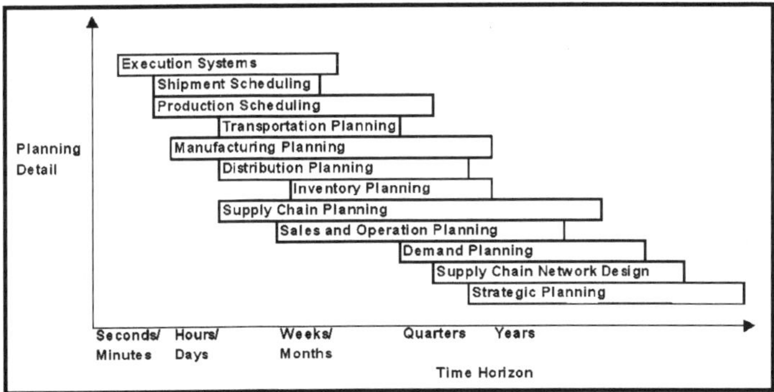

Fig. APS solutions related to the time horizon (Bermudez, 1998; revised)

DIFFERENCES IN PLANNING HORIZONS

The enumerated solutions can roughly be divided into three levels of planning and scheduling:

- Supply Chain Planning

- Manufacturing Planning
- Production Scheduling

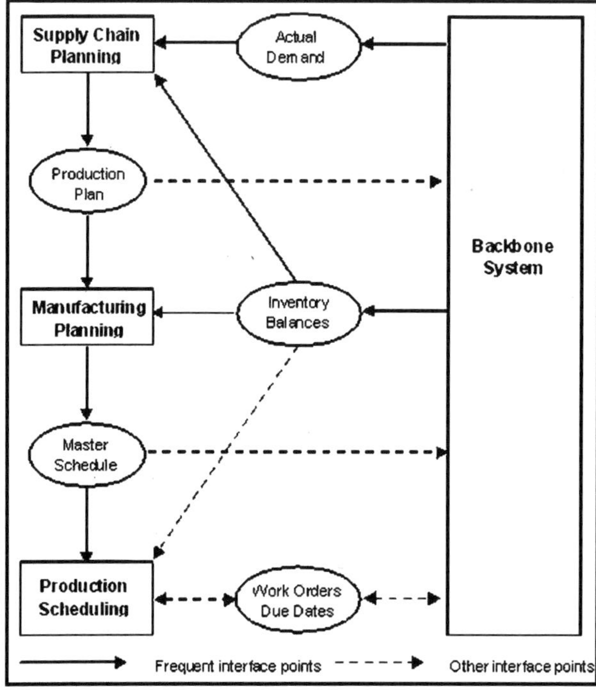

Fig. Relationships of major planning functions with typical data flows (Bermudez, 1998; revised).

The first two levels can be called planning-centric systems. These systems focus on long term strategic and some tactical objectives. For a global or a multi-site company, these systems can optimise the best possible location in a network of manufacturing locations where a specific order must be produced.

The planner enters the business objectives into the system, after which the planning engine determines which objectives might be violated. When objectives are violated in the long term it is possible to adjust the constraints, which results in gained objectives. Adjustments in the constraints might be possible if there is enough time. When there is not enough capacity, in the long term this constraint can be eliminated, because capacity can be enlarged by acquiring an extra production line (Hess, 1998).

The third level is more a scheduling-centric system. These systems focus more on operational and some tactical objectives. The task of a production scheduling system is to generate a feasible production schedule given a required production output. The constraints it deals with are quite real, they are often given and allow only limited changes (Hess, 1998).

Supply Chain Planning

This SCP group takes a forecast and looks at actual demand, after which a constrained operation plan for both manufacturing and distribution is generated. A multi-plant constrained master schedule, is the output of the SCP process for manufacturing. To create this output the material availability's and plant capacities are accumulated. For some industries, transportation requirements and set-up sequencing are considered as well.

Advanced Manufacturing Research (AMR) describes SCP as follows (Bermudez, 1998):

"SCP determines what should be made given the available resources to achieve business goals."

Manufacturing Planning

The output from manufacturing planning generally is a constrained master schedule for a single plant or a group of similar plants. This master schedule considers the constraints in a more detailed perspective than in SCP. In manufacturing planning a full MRP I/II explosion can be included in the process.

AMR describes Manufacturing Planning as follows (Bermudez, 1998):

"Manufacturing Planning determines how and when it should be made based on material and resource constraints to meet customer demand."

Production Scheduling

The goal of this group is to translate the output of the supply chain planning to an operational plan and work orders. Here is where the ultimate specification takes place on the basis of which the suppliers will deliver., the production departments produce and distribution receives and ships the products. APS supports the planner by continuously adapt or suggest adaptation of the planning and scheduling based on the recent information. Product scheduling is designed to produce the most efficient production schedule (where the throughput times are minimal, the output maximal and the costs are low).

PLANNING AND SCHEDULING

An APS system uses the following planning and scheduling approach: A planner module which pays some attention to capacity constraints produces a "scheduleable"plan. This plan then feeds a scheduler module, which produces a detailed list of operations showing how capacity will be used and returns this information to the planning function for use in the next planning period. The data regarding current and planned operations can also be used to provide realistic estimates of the ability to meet a new customer order request. This integration of planning and scheduling is described in the following two paragraphs.

Advanced Planning

The role of planning in APS is to determine what demands on the production system will be met over a given planning horizon. The input to the planning process includes information on manufacturing capacity and demand data. Demands may be of several types: customer orders, forecast, transfer orders (i.e., orders from other plants), released jobs, or replenishments of safety stock. Manufacturing system data includes bills of material, workcenter availability, part routings through workcenters, and inventory (both on-hand and scheduled for delivery). The output from the planning process is a feasible plan, which provides release and completion times for every demand. Like MRP, APS takes into account the availability of materials. Unlike MRP, it also takes into account the capacity of workcenters to process the material and satisfy demands. This planning process is order-centric, focusing on the demand for end items and determining how much demand can be met in a given time period. Exactly how that demand will be met, in terms of specific assignments of jobs to workcenters and their sequencing, is left to the scheduling function. It is in fact often desirable for a plan to be somewhat tentative, since it covers a planning horizon subject to disruptions. Forecast may not be accurate. Deliveries may be delayed. Equipment may fail. Unexpected rush orders may be received. Therefore planning is not expected to be highly detailed. Individual machines may be aggregated into a workcenter with no determination of which will be used by a specific order. Setup times may be averaged since sequencing at this time is premature. Buffer times may be defined, especially prior to processing on bottleneck machines, to allow for possible disruptions. The end result is a "scheduleable"plan.

Advanced Scheduling

The role of the scheduler module in APS is to produce a detailed list of operations specifying which orders are to be worked on at which workcenters and at what times. The input to this module includes all demands to be satisfied, including the internal orders added by the planner module when an end item required a component to be manufactured. It includes the current material inventory levels as well as planned deliveries or purchased materials. It also includes the same manufacturing system data as that provided to the planner module but uses a more detailed representation of that data. Detailed information used by the scheduler module that is not pertinent to the planner module includes:

- Variable run times based on the machine and operator actually assigned.
- Rules for selecting machines and operators based on skill sets and quality requirements.
- Variable setup times based on the previous and next part characteristics such as part type, family, colour, width, etc.
- Rules for sequencing jobs at workcenters, based on minimising setup and other factors.

- Allowable shift overruns.
- Rules for selecting from a list of prioritised jobs based on due date, slack, cost and other factors.

The result is an accurate representation of what to expect on the shop floor in the immediate future. While the planner module typically considers demand on the system over a few weeks or months, the scheduler module will typically work with a much shorter time frame such as a shift, a day, or a week. The usefulness of a detailed schedule degenerates quickly as time passes, since disruptions on the shop floor or changes to the order mix may require significant adjustments.

FEATURES OF APS

An APS system has a number of features that enable it to be clearly differentiated from traditional planning systems such as MRP I/II and DRP.

Concurrent planning

In the traditional planning process, as in the case of MRP I/II and DRP, three main variables can be distinguished:

- demand
- materials (raw material and semi-manufactured articles)
- capacity.

The traditional planning process is the so-called 'waterfall approach', in which the planning process is undertaken sequentially. It starts with an MPS, after which MRP I/II and CRP are performed. The sequential approach decouples the plans from each other and cohesion can only be preserved by constantly repeating the planning process. In the traditional systems production is based on a plan that is already outdated, since there are new orders and other changes. In case of 'concurrent planning', however, the three main variables are considered simultaneously. This results in synchronised, optimal planning for the chain as a whole, based on the most up-to-date data. It should be noted in this context that APS uses certain core data, such as the capacity per production location and certain core constraints, which are mentioned below. The two planning processes are described in figure. Chapter 5 discusses this functionality more thoroughly.

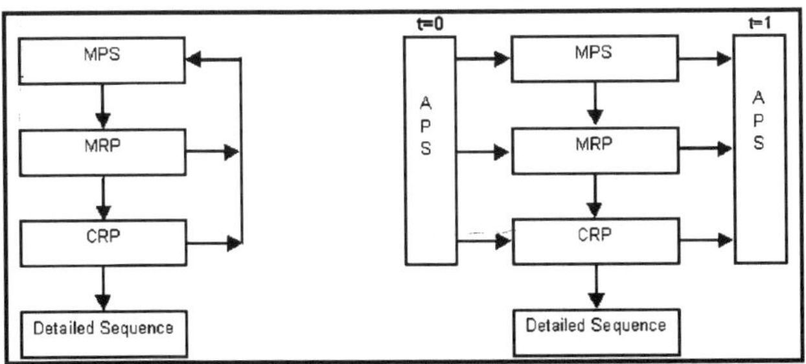

Fig. The traditional planning process.

Constraint-based planning

A second important characteristic of APS systems is that account is taken of the constraints present in an enterprise, such as capacity and materials. APS systems use these constraints to model the production and distribution environment. The performance that an enterprise can achieve is determined by the constraints.

Various constraints can be identified (Bermudez, 1998):

- Material availability
- Available capacity
- Enterprise policy
- Cost
- Distribution requirements
- Sequencing for set-up efficiency

Speed

The speed of planning is an important characteristic. Improvements in computer processing power and software design has lead to good response times. As a result, a customer can be informed about the delivery possibilities within a few seconds. The person in contact with a customer who wishes to place an order has a strong negotiation position since he has a picture of the possibilities that the company can offer the customer. If the company is not able to satisfy the customer's wishes, he is immediately able to offer alternatives to the customer. Speed is also important during the planning cycle. Since all the links in the chain are now closely co-ordinated, delays in one link can have an amplified effect in the subsequent links.

Preferences

It is possible to indicate preferences in APS for purposes of strategic decision making. It is possible to regard certain customers as strategically important. In APS this is interpreted as a customer with a higher priority. These strategic customers must be considered as such throughout the whole organisation. This avoids a situation in which one sales organisation regards a particular customer as strategic, while for another sales organisation the same customer is unimportant.

It is also possible to allocate priorities to products. For a manufacturer of compact discs, for instance, it is highly important for singles never to be out-of-stock. These singles are therefore allocated a higher priority than albums, for which an out-of-stock situation is less damaging.

What-if simulation

One of the first, and still most common applications for advanced planning and scheduling products, is decision support using the facility for what-if simulation. It is possible for various alternatives to be entered into the system and for the system to maximise company profit and/or minimise costs, subject to the condition that the order can be delivered on the date required by the customer. The planner can examine various scenarios under which the order is delivered and the system subsequently indicates the consequences of the various scenarios for existing orders. A graphical interface makes it easy for the planner to compare the various alternatives computed by the system, so that the most acceptable solution can then be chosen. The planner can 'play around' with the data, with the most acceptable alternative being chosen and used as new input.

While all APS products can be used for simulation and what-if analysis, some vendors provide more complete facilities to compare plans and schedules. This ranges from the ability to have multiple copies of different plans visible for die-by-side comparison (such as ERP systems) to the ability to produce cost analyses of various planning options.

The Advanced Manufacturing Research Inc. (AMR) believes that the potential of advanced planning and scheduling for widespread management decision support has not yet been realised. Generally, decision support is limited in scope to tactical manufacturing operations, such as introducing a new product or accepting a large order. APS also has the potential to support strategic management decisions, such as adding or dropping new plants, combining operations, and testing the impact of marketing promotions. Currently, extensive training is often required to do this level of simulation. This limits its use as a decision support tool to a few "power users." Some decisions such as closing a plant, may be too sensitive for anyone but senior management. Several vendors are working on improvements to their modelling capabilities and user interfaces to enable managers to make more extensive use of the decision support aspects of APS systems for enhancing general business planning.

Available to Promise (ATP)

APS can be used to obtain a better insight into ATP. ATP represents a rolling balance of "unconsumed supply" (uncommitted portion of the inventory) over time. "Unconsumed supply" is inventory on hand, plus planned supply, minus existing commitments to customers. The ATP allows a company to see what inventory has not yet been allocated and what can be done with that inventory for potential customers in a specific period. The planner is enabled to adjust the input and the presented solutions using his own know-how.

When an ATP function receives an order, it slots the order for the day (or days) on which there is sufficient supply available to cover the order quantity. Based on the slotting dates, the function proposes a delivery date (or dates) to the customer. By having insight in the organisation, an order-taker can check availability throughout the organisation. Due to insight into the organisation, the order-taker can give the customer delivery options. The customer can, for example, choose between road transportation or air transportation, which is more expensive but faster (McKenna, 1998).

The table below illustrates a printed circuit board (PCB) manufacturer's planned supply, committed orders, and the resulting product availability (ATP) for a particular product:

Table. An example of ATP

	Beg. Inv.	Period 1	Period 2	Period 3	Period 4
Planned Supply	100	600	800	1000	1000
Committed Orders		500	800	900	800
ATP		200	200	300	500

The manufacturer begins with 100 PCBs in inventory and plans to produce 600 PCBs in Period 1, 800 in Period 2, 1000 in Period 3 and 1000 in Period 4. The manufacturer has committed to delivering orders totalling 500, 800, 900 and 800 PCBs in Periods 1 through 4, respectively. As a result, the manufacturer has 200, 200, 300 and 500 PCBs available to promise to incoming in Periods 1 through 4.

Suppose a customer, who only accepts shipments in lot sizes of at least 200 PCBs, places an order for 500 PCBs in Period 2. The PCB manufacturer could promise 200 PCBs in Period 2 and 300 units in Period 4. However, the PCB manufacturer would not be able to promise 200 PCBs in Period 2, 100 in Period 3, and 200 in Period 4 because the customer's minimum lot size is 200 PCBs would be violated in Period 3.

There is a difficulty in performing ATP by simply committing 300 PCBs to an order in Period 3 and then considering PCB availability in Periods 1 and 2. Availability in Periods 1 and 2 should drop to zero so that the manufacturer can respect the commitment in Period 3. Order promising thus impacts availability both on the days preceding and following the days on which orders are slotted. Further, availability is impacted just as much when customers cancel orders as when they place them. Finally, suppliers rarely offer only one product. More often, they offer numerous products, some of which are interdependent from an order promising perspective (for example, a customer only wants a CPU if it is shipped with a monitor). In an environment requiring reliable, real-time response, performing ATP manually is simply not an option.

Capable to Promise (CTP)

The next step after ATP is capable to promise. CTP integrates order promising and supply chain planning. Now, the order-taker does not only look at the uncommitted available stock, but also production capacity and material availability are taken into consideration (McKenna, 1998). CTP derives from the real-time APS engine a delivery date by adding a customer order in the system, where after this engine determines when the order is scheduled to be produced, by looking at available material and capacity (Bermudez).

If an ATP query determines that available supply is insufficient to cover a particular order, the CTP supply chain planning function enables the supplier to exploit capacity and material opportunities, if any, to increase planned supply in time to accommodate the order. If, in the previous example, an Original Equipment Manufacturer (OEM) were to order 300 printed circuit boards to be delivered by the end of Period 2, the manufacturer would not be able to promise the order on time, based on the planned supply used in computing ATP. However, CTP would automatically access the feasibility of increasing the planned supply in Periods 1 and/or 2.

While most manufacturers like the CTP concept, they often have trouble envisioning its application in their company. The idea that the customer service or order-processing department would, in effect, be scheduling the plant is too radical, if not logistically impossible for most manufacturers. In spite of the emotional response, the AMR believes that the CTP concept is fundamentally sound. This technology offers substantial benefits which will resolve the organisational issues.

Profitable to Promise (PTP)

ATP and CTP only look at the possibility to deliver the order on time to the customer. It would be better to be able to accept the order based on the financial implications for the company. This is called profitable to promise. The implication of this step might be that an order is rejected today, because now the capacity can be left available to a future unrealised order which is more profitable. With PTP you can assure that the right customer gets the right order at the right time, which is most profitable to the organisation (McKenna, 1998).

Bi/multi-directional change propagation

Changes occurring in the production process, such as breakdown of a machine in a production line, are reported immediately to the APS system. The planner can then adjust the planned activities upstream as well downstream using APS. This is referred to as bi-directional change propagation.

The threat that, as a result of the breakdown of the machine, the enterprise will be unable to deliver certain orders on time. The system now presents solutions, for example allocating the orders to another production line, and/ or using unused but operational machines in the line for other orders or parts of orders. As a result, capacity continues to be used optimally and customer service remains high. These solutions are an example of multi-directional change propagation. Bi/multi-directional change propagation is particularly used in scheduling-centric APS systems.

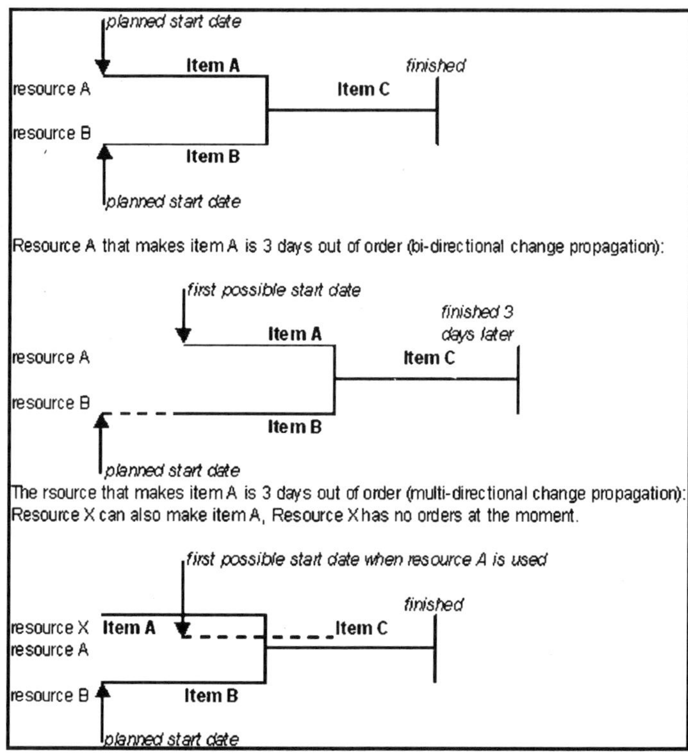

Fig. Example of bi/multi-directional change propagation

Bucketless planning

In the case of traditional planning methods the planning process uses 'time buckets' with a schedule being drawn up for a specific period. In scheduling-centric APS, planning in terms of time buckets is abandoned and continuous short-term planning is undertaken. Planning is undertaken as far as possible on the basis of actual orders rather than forecasts. Planning for the medium and short term continues to be undertaken in terms of buckets.

Reliability

This is the possibility of making promises concerning delivery times and delivery dates and also fulfilling such promises. It is possible to inform the customer of the ultimate delivery date. When the customer places his order, the company gives the delivery date and has the possibilities to adhere to that promised date.

Chain approach

Considering the entire chain simultaneously makes the chain more transparent. The planner can use graphical interfaces to visualise the entire chain and drill down into these chain parts to look closer at possible problems that occur. The planner can, for example, when a specific order cannot be produced drill down into the production system to look at the machine experiencing a capacity problem. The planner can alter the schedule to solve this problem, for example by rescheduling the orders regarding the machine.

Optimisation

Optimisation means generating the best solution to a specific problem (Proasis, 1998). APS can be used to optimise both tactical and strategic business issues. At the tactical level the system can help to optimise sourcing, production and distribution plans. At strategic level APS supports in optimising the network configuration (Bendiner, 1998). Different techniques can be used to solve the optimisation problems (Bermudez, 1998):

- Linear Programming
- Genetic Programming
- Theory of constraints
- Heuristics.

Alternate Routings

An APS system is able to check all possible production routings to optimise the production schedule. Traditional planning systems work with preferred supplier routings, which means that for all product combinations fixed routings are entered into the system. Customer A, for example, receives his order always from DC "X". With alternate routings it is possible, if DC "X" is not able to meet customer due dates, to check the possibilities of delivering from another DC, which has available capacity to deliver the order on time to customer A.

Total Order Management (TOM)

APS systems can be used for TOM. This means it can be used as the central and critical function of the organisation. To collect all the needed information to optimise plans an APS system make use of intelligent client processes (ICP). These processes act as intelligent agents, that collect all the information that is needed for the planning engine to make decisions. An example will illustrate the TOM process.

As soon as an order is entered into the APS system, the appropriate intelligent agents will check availability of components. Each ICP will return a delivery schedule for the needed components with associated costs. Together with this information and the capacity information a delivery schedule is produced. Based on this delivery schedule a pricing ICP will deliver the associated prices for each order. The TOM process includes all the processes from order entry to shipping (Hadavi, 1998).

APS IN RELATION TO TRADITIONAL PLANNING SYSTEMS

The traditional planning systems like MRP I/II and ERP are not optimal. In this chapter the differences between these older traditional systems and APS will be explained. In the first paragraph MRP I/II will be compared with APS. In the second and last paragraph ERP and APS will be compared.

APS versus MRP I/II

There are few assumptions underlying MRP I/II, which do not apply for APS (Turbide, 1998):

- All customers, product, and materials are of equal importance. In an APS system preferences can be inserted into the system, which means that for example some customers are more important than other customers.
- Lead times are fixed and known. With APS it is possible to reduces lead times, because the system is able to contact suppliers to get materials earlier (at a higher price).
- It is a top-down, one-pass, sequential process. With APS it is possible to adjust schemes in a multi-directional way.

Other disadvantages of MRP I/II are:

- MRP I/II runs are batch-oriented and take hours to complete. Because it is a time consuming process, it can only be done at night or in the weekend (Turbide, 1999). When you want to adjust the schedule, you have to wait for the next day to see if the adjustment turned out well. When an adjustment in a plan or schedule has been made, the APS system recalculates the plan or schedule within a few seconds or minutes
- MRP I/II does not give any possibilities for decision support or simulation (Turbide, 1999). APS has the ability to perform a what-if analysis. Different scenarios can be compared with each other and the best one can be filed into the transactional system.
- MRP I/II systems deliver long reports that force the end-user to dig through the details to find the problems. APS systems are easy to learn and they work with exceptions. When an exception occurs, the system reports a problem and the user-friendly interfaces allow the user to drill down into the specifications to identify where the problems occur. When the problem has been identified it is easy to administer solutions into the system (Grackin, 1998).
- The material allocation in MRP I/II is done on a first-come-first-served basis. This can result in plans that are suboptimal (Bermudez, 1998). For example, you have 25 units in stock and there are two customers ordering this unit. Customer A is first and wants 50 units and customer B wants 25 units. Because customer A is the first the 25 units in stock are reserved for this customer and 50 units are

scheduled to be produced. Both customer A and B have to wait until these units are produced and are unsatisfied with the delivery times. An APS system deals with this problem in another way. It allocates the 25 units in stock to customer B and starts the production of the 50 units for customer A. At least customer B is satisfied now, because he receives his units at once.

APS versus ERP

ERP systems are very strong on transaction processing and execution of standard repetitive tasks, but their true planning and decision support capabilities are very limited, and as a result, frequently fail to deliver their full potential (Proasis, 1999).

There are a number of reasons why ERP systems failed to improve manufacturing planning (Bermudez, 1998):

- The level of detail in ERP systems is too rough for adequate decision making. Also, the existing technology which is used for ERP systems does not allow greater detail for real time analysis and simulation, which enables adequate decision-making.
- The tools used within ERP systems are used infrequently and are sometimes incomprehensible for senior management.
- There is no consideration given to the interdependency of material and capacity availability.
- Multi-plant planning at one time is not possible.
- Actual results are not entered into the system to make process and data improvements.
- Optimisation of the production schedule to improve throughput is not possible.
- The lead times are not dynamically calculated but static and manually assigned.

All these named points are disadvantages of ERP systems. APS systems are able to do all these things. For example, APS systems can do multi-site planning at one time.

ERP systems are designed as a suite of applications around a database, which means that applications communicate with each other via the central database. The disadvantage of this procedure is an iterative procedure of going back and forth between applications, which make the transaction update time very long. As a result it is not possible to give real-time response to customer enquiries. An other disadvantage is that customer constraints or preferences cannot be dealt with in an easy way. APS systems, on the other hand use an integrated environment. The logic of the order entry is part of the logic of the planning and scheduling engine. In an integrated environment, the planning and scheduling engine will follow all "rules and preferences" before an answer to the customers inquiry will be given. Some examples of these "rules and preferences" are: 90% of product group S must be shipped on time, or all products for customer B must be shipped together (Hadavi, 1998).

APS FOR PRODUCTION ORGANISATIONS

APS has specific possibilities for producers. When implementing an APS system it is also possible to have APS-systems running on factory level (per production location). At this level the system optimises the production location, given the orders from the central APS system. The local running APS-systems are connected to the central APS that works on the whole chain. At this level the scheduling comes in. As described in paragraph 4.2 the difference between planning and scheduling is not always clear. Planning concerns the overall picture and focuses on the longer term, while scheduling focuses on the individual orders that have to be processed in succession with more specific constraints.

APS FOR DISTRIBUTION ORGANISATIONS

As yet APS has found use in production organisations. Important uses are found in the semi-conductor industry. These products know a large amount of production stages. These stages can be performed in large-scale production centres around the world. The optimisation of the flow of goods and the capacity over all the location, is an absolute necessity for the organisations in this industry. For distributors (retailers, wholesalers, distribution organisations) the use of APS is not so obvious. The reasons for optimising the supply chain can not be found in the optimal use of capacity or price control, but manly in the maximisation of the product availability and the optimisation of the stocks. This asks for a good planning of the future demand (demand planning or sales and operational planning) and the almost continuous registration of the real demand and available stocks in the supply chain. For a distributor the modules for demand planning (available-to-promise, distribution- and transport planning) are the most important.

COMPUTER AIDED PROCESS PLANNING

In Machinery Industry

PROCESS PLANNING AND PROCESS PLAN

Definitions of process planning

Process planning can be defined as engineering activities assigned to manufacture or assembly the products. Process Planning is the activity in a production enterprise that determines which processes, materials, and instructions will be used to fabricate a product. Process planning dictates a manufacturing facility, processes and parameters which are to be used to convert materials from an initial form to a predetermined final form.

Process planning is an act of preparing detailed process instructions to produce a part. Process planning is a result of engineering planning activities of process planners which prepare a list of processes needed to convert a raw material shape as a starting point into a predetermined final shape.

The aim the of process planning is to convert design specification into manufacturing instructions and to make products within the function and quality specification at the lowest costs. Process planning is a recipe for making products. Process planning in manufacturing may be generally defined as the development of set of instructions describing all the operations required to convert a design into a product.

PROCESS PLAN AND PROCESS PLANNING

Engineering parts are manufactured by some of engineering technologies – machining, forming, casting or welding. Production is carried out according to prescribed process instructions and operations. The prescription is stated in a process plan. The process plan is documentation determining all machining operations, order of the operations, all machine equipment, conditions and requirements to producing a part. Process plan consists of preparing a detailed plan with instructions specifying how a product is to be manufactured or assembled, taking into account the characteristics of both the product and the available equipment and processes.

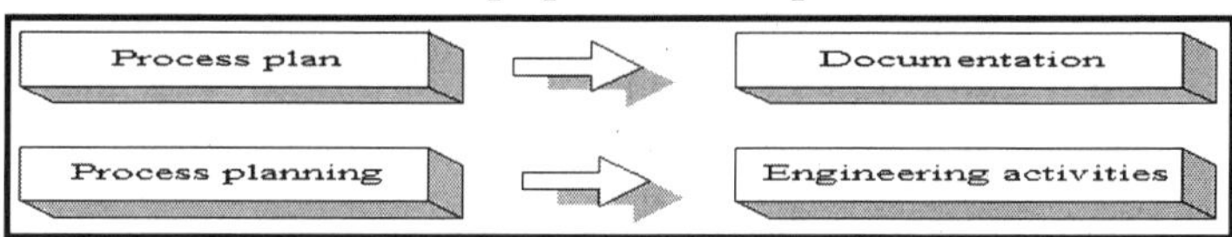

Fig. Process plan and process planning

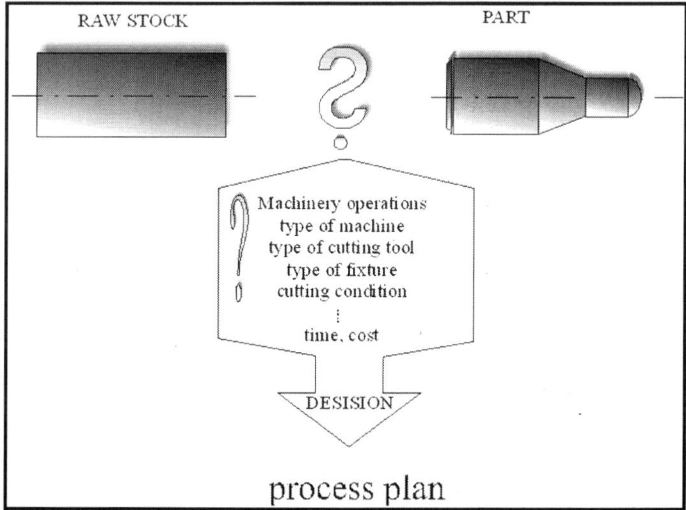

Fig. Process plan as a recipe for transformation of raw stock to the final part.

PURPOSE OF PROCESS PLANNING

Process planning deals with the selection of the processes and the determination of conditions of the processes. The selected operations and conditions have to be realised in order to transform raw material into a given shape. All the specifications and conditions of operations are included in the process plan. The process plan is a document such as engineering drawing. Both the engineering drawing and the process plan present the basic document for the manufacturing of products. Process planning meaningly influences time to market and productions cost. Therefore the planning activities have a great meaning for competitive advantage.

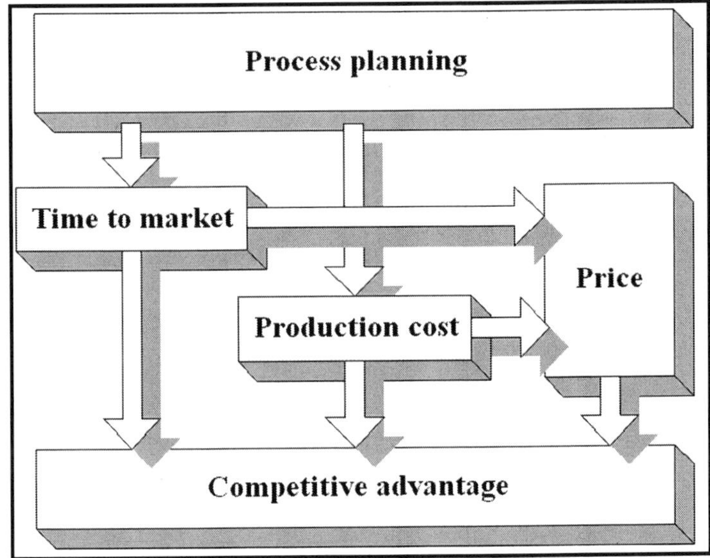

Fig. Influence of process planning on the competitive advantage

Process planning is the planning activity of an engineer or a planner. One can see planning activities in various application areas, for example:

- Food industry,
- Chemistry industry,
- Building industry
- Engineering industry.

In the engineering industry following activities for various technologies are planned:

- Machining,
- Forming,
- Casting,
- Welding,
- Heat treating,
- Assembling.

The product is manufacturable in many ways. Manufacturing methods depend on several parameters of part and production:

- Geometrical properties (shape and dimensions) of part,
- Material properties of part,
- Total amount of parts,
- Batch,
- Available manufacturing equipment,
- Production facilities,
- Constraints of manufacturing environment,
- Time to manufacture,
- Cost to manufacture.

A human has to consider the mentioned parameter and to propose an optimal process plan. The optimal process plan has to fulfil the selected production criteria (cost, time, quality). The manufacturing environment and the production facilities are subjected to relatively small changes over time. This means that processes, machine tools, cutting tools, measuring machines, fixtures, jigs and other equipment are not subjected to great changes over time. However, the planner must take into account the constraints of the manufacturing environment.

During process planning a planner operates with:

- Information located in engineering drawings and bills of material,
- Manufacturing knowledge and
- Information of machine equipment.

Process planning is an engineering activity with a great amount of decisions. It calls for good knowledge and experience on the part of the planner participating in the process plan making.

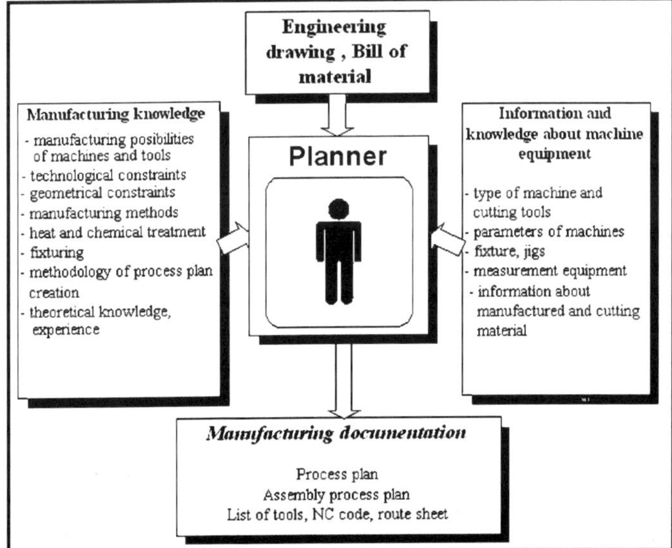

Fig. Information and knowledge needful for process plan creation

The process plan is realised step by step in individual planning tasks. According to the process planning there are multilevel planning activities. Into output of a previous task influences the following planning tasks. Individual planning steps are shown in the following picture.

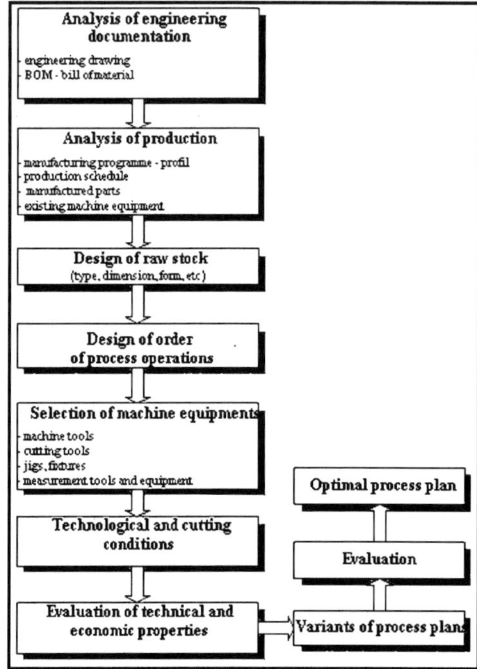

Fig. Individual tasks of the process planning

MANUFACTURING PROCESS PLANNING

In a scientific literature the term - process planning is also known as

- Manufacturing planning,
- Material processing,
- Process engineering,
- Machine routing.

Process planning in manufacturing may include the following activities:

- Selection of raw-stock,
- Determination of machining methods,
- Selection of machine tools,
- Selection of cutting tools,
- Selection or design of fixtures and jigs,
- Determination of set-up,
- Determination of machining sequences,
- Calculations or determination of cutting conditions,
- Calculation and planning of tool paths,
- Processing the process plan.

The level of development of process documentation depends on production type (piece, batch or mass production). If the product is to be manufactured on Numerical Controlled (NC) machines, process planning includes also the generation of NC programs. The degree of detail of a process plan varies from industry to industry. The job-shop type of manufacturing environment usually requires the most detailed process plans since the design of tools, jigs and fixtures and manufacturing sequence etc., are dictated directly by the process plan.

PROCESS PLAN AS A SET OF MANUFACTURING FEATURES

The process plan involves process operations. The process operation is possible to consider as a manufacturing feature involving the machine, cutting tools and fixture. Subsequently the process plan is a set of the manufacturing features. It is important to determine the condition of utilising the machine equipment and environment. The cutting condition (cutting speed, feed and depth of cut) represents significant condition for machine equipment.

MACRO AND MICRO PROCESS PLANNING

There is methodological possibility to segment process planning activity in two following section:

- Macro process planning,
- Micro process planning.

Output of macro process planning activities is a scheme of machine equipment and process operation sequencing. Micro process planning includes determination of cutting condition.

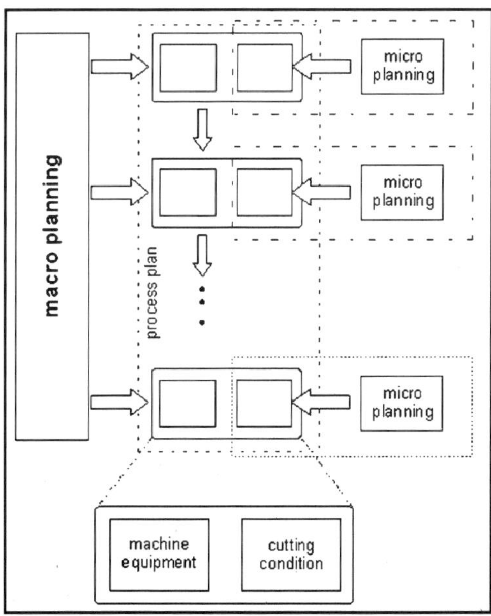

Fig. Micro and macro process planning

Macro Process Planning includes:

- Review of common machining processes,
- Accessibility of manufacturing features,
- Process and machine selection,
- Operations sequencing methods,
- Workpiece set-up planning.

Micro Process Planning are activities as following:

- Advanced process models for prediction of machining performance
- Process optimization,
- Optimum tool path planning algorithms,
- Analysis of Part-process-workholding design trade-off.

Oftentimes process planning is accepted as a manufacturing recipe with specified manufacturing equipment and conditions. However, the process plan is necessary to understand as a complex task to transform raw-stock not only to final product but to the customer. It comes to this, that in the complex process plan the following problems are involved:

- Machining,
- Transportation,
- Assembly,
- Metrology,
- Unit and system test,
- Packaging and
- Export.

Recently, and also often nowadays, experts have mainly performed these tasks. This work involves both creativity and skill. Many of activities are routine based. When creativity and skill are not sufficient to solve a process planning problem, the process planner will usually communicate the problem to the designer in order to try to solve the problem. Therefore, process planning is a team job which includs experts from the individual application area (tool designer, fixture designer, process planner, cutting condition planner, etc.).

Communication among the process planning team has great balance, a good process plan often depends on joint co-operation and team-work.

During the process planning the following human abilities play an important role:

- Theoretical knowledge,
- Skill,
- Experience.

Goals of process planning are:

- Meet quality requirements,
- Maximize production efficiency,
- Minimize production cost,
- Protect working conditions,
- Improve technology.

The process planning job, except the basic task of plan generation, should check design feasibility and also generate cost estimates and bid product information.

INPUT AND OUTPUT OF PROCESS PLANNING

It is possible to consider process planning as a closed subsystem with input and output. Following basic input parameters for the process plan generation are:

- Manufacturing resources,
- Engineering specifications,
- Design drawings,
- Bills of material (BOM).

Output from the process planning is a set of the following:

- Material and blank specifications,
- Enterprise and machine routings,
- Machine settings,
- Operating instructions for each manufacturing task,
- Measurement equipment and measuring activities,
- Tool and fixture descriptions,
- Required process capabilities and
- Time and cost estimates.

TYPE OF PROCESS PLAN

The type of the process plan depends on the production type. The same part may have an other process plan in a different enterprise. A different process plan belongs to the same part in piece, batch and mass production.

The level of process plan elaboration depends above all on the type of the production (piece, batch and mass).

There are several varieties of process plans:

- Textual process plan,
- Pictographic process plan,
- NC code,
- Simulation code.

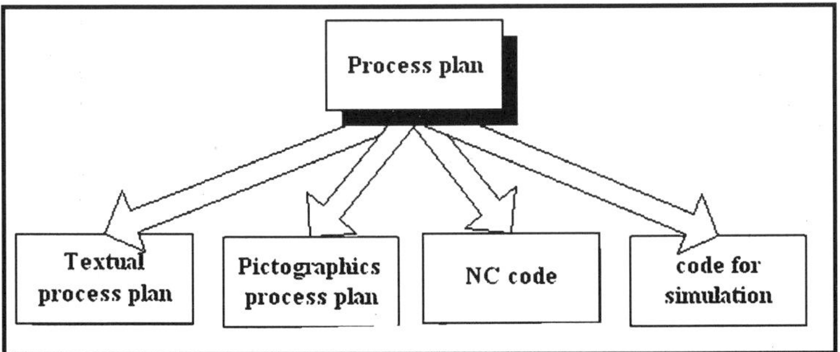

Fig. Types of external representation of process plan

COMPUTER SUPPORT AT PROCESS PLANNING ACTIVITIES

Why computer support for process planning?

Competitive surroundings and environment force manufacturing enterprises to produce in smaller series, with shorter lead time and with decreasing costs. Flexibility is one of the keys to solve the current state of production.

Process plan includes standard process operations. The process planning task is to retrieve and to synchronise these operations into an effective and optimal process plan. Synchronising of the process operation is a high mental activity for the process planner. As process planning is a complex task, it is needful to take into consideration the affecting factors. Parallel reflection and reasoning is a characteristic feature of the human.

Possible type and range of activities in process planning:

- Intuitive activities (from 1 to 5 %),
- Formal logical activities (from 25 to 50 %),
- Routine activities (from 45 to 74 %).

The routine activities are solved with known algorithms. A more complicated matter is the computer solving of problems with unknown algorithms (e.g. selection of acceptable machines and tools, selection of operation sequence, etc.). Definite algorithm for some specification of the process plan solving is not known in advance. The process planner often utilizes heuristic solving methods, knowledge, experience and intuition. These activities have been practiced for many years and knowledge of research in the area is often of great importance in the decision process.

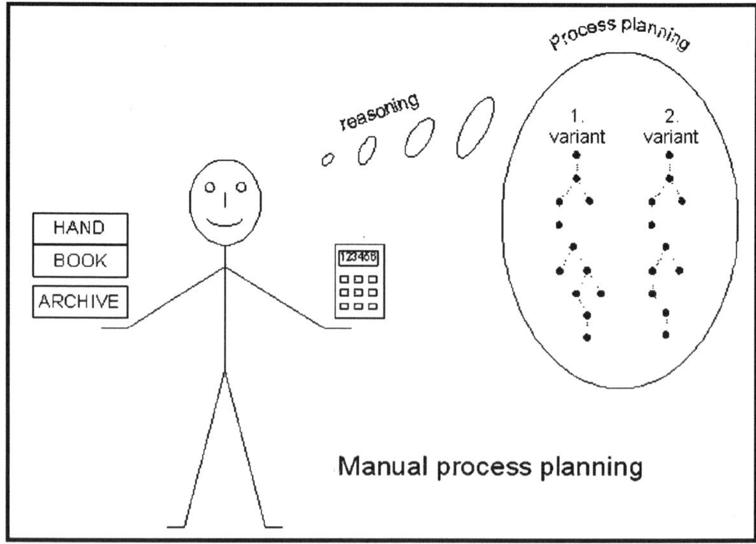

Fig. Manual process planning

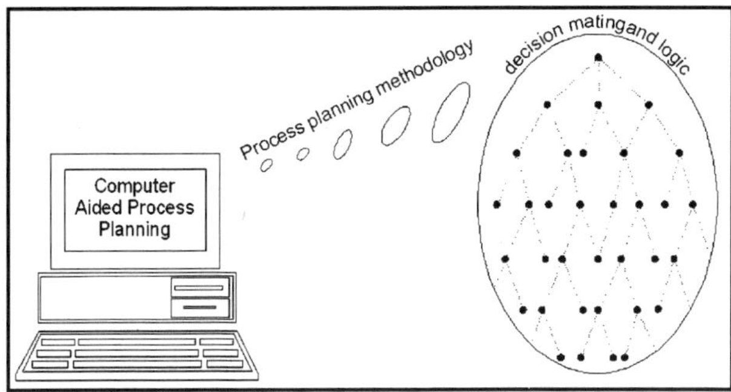

Fig. Computer support in process planning

The manual process planning is often a tedious and time demanding engineering process and it is one of the labour activities in the preparatory stage of manufacturing. There are many routine, heuristic, deciding and intuitive activities used by a planner. There is an effort for these activities to be supported by computer. Computer support can markedly help to solve some planning activities. Computer aided process planning (CAPP) system is software for the automated design of route sheet. A number of reasons can be identified for the advance of CAPP systems. There are amount of process planning departments in small batches where a skilled workforce is scarce. Many companies have different process planners make different process plans for the same parts, resulting in inconsistencies and extra paper work. CAPP systems can help in overcoming these inconsistencies.

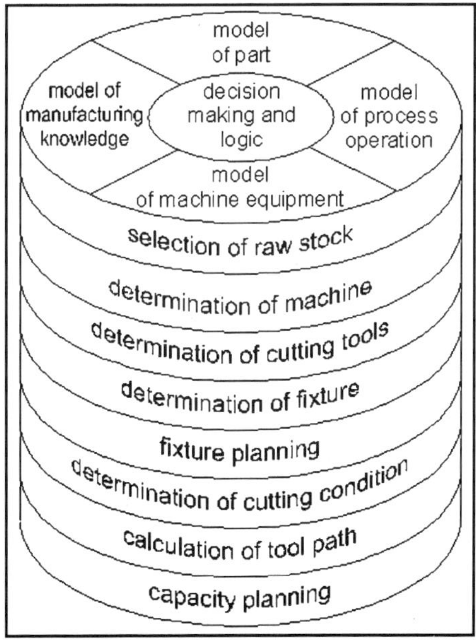

Fig. Computer aided process planning

Computer Aided Process Planning aids in creation of process plans for manufacturing and increases the flexibility of manufacturing. Process planning is a task which requires a significant amount of both time and experience. Computer support or computerised process planning systems can help reduce a process planning time and increase plan consistency and efficiency.

BENEFITS OF COMPUTER-AIDED PROCESS PLANNING

The use of CAPP systems has the following potential advantages:

- Reduced demand on the skilled planner,
- Reduced process planning time,
- Reduced process planning and manufacturing cost,
- Created more consistent plans,
- Produced accurate plans,
- Increased productivity,
- Increased high flexibility,
- Attained high efficiency,
- Attained adequate high product quality,
- Possibility of integration with the other automated functions and systems.

Methods of automated manufacturing process planning

The advantage of automated manufacturing process planning is undisputed. There are two approaches for creation and processing of process plan based on computer support and advanced planning methods. The first approach is based on Group technology utilising (variant method), the second approach is the exact mathematical principle based on modelling of part, manufacturing knowledge and process plan (generative method).

In the history of computer aided process planning two different ways of obtaining the process plan can be observed:

- Variant process planning,
- Generative process planning.

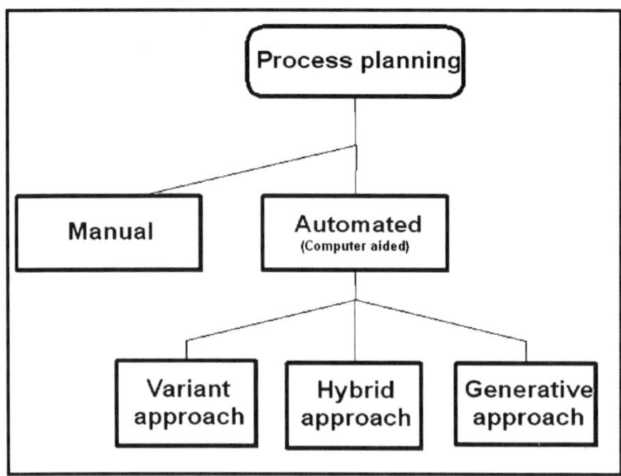

Fig. Kind of process planning

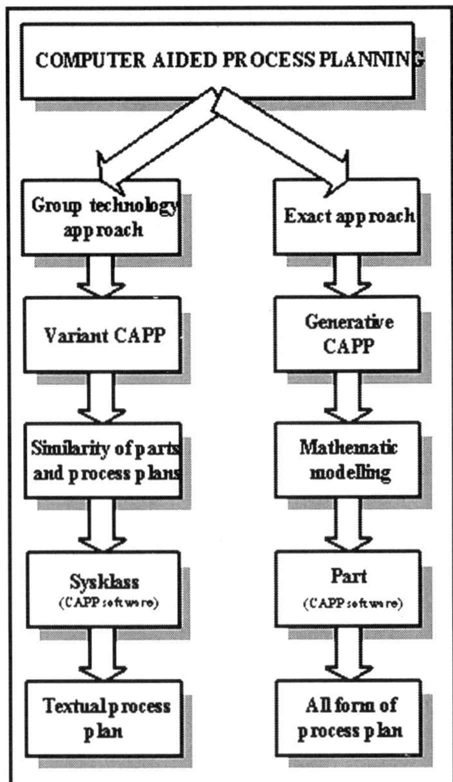

Fig. Two basic approaches for CAPP

Variant Process Planning

Variant process planning implements a coding and classification scheme by which a process plan for a previously planned part is retrieved. The retrieved plan is based on the similarity to the new part. The process plan is then manually modified as required for the new part design. There is high probability that similar parts have similar process plans. This is a basic assumption of utilising the variant process planning method.

Variant methods assume that the user is able to determine the appropriate classification codes needed to retrieve appropriate plans, and that plans exist and include features which are closely analogous to those of the new part. Because variant systems are based on the process plans on historic data, it is also assumed that the factory configuration is stable, with only minimal workstation or process capability changes.

The variant approach to CAPP was the first approach used in computer process planning. Variant CAPP is based on the concept that similar parts have similar process plans. The computer is used as a tool to assist in identifying similar process plans, as well as in retrieving and editing the plans to suit the requirements for specific parts. Variant CAPP is related to part classification and Group Technology coding. In these approaches, parts are classified and coded based upon several characteristics or attributes. A Group Technology code can be used for the retrieval of process plans for similar parts.

Generative Process Planning

The process plan is created according to a very complicated methodology. The description of part and knowledge bases are necessary for the process plan creation. In a generative process planning system, an adequate (usually feature-based) part model is utilised to build a process plan from scratch. From the fully described geometry, parameters are derived to determine applicable processes and resources. Generative process planning allows a high level of flexibility for building the process plan according to varying resource conditions.

Generative CAPP came into development in the late seventies. It aims at the automatic generation of process plans, starting from scratch for every new workpiece description. Often, the workpiece description is a CAD solid model, as this is an unambiguous product model. A manufacturing database, decision making logics and algorithms are the main ingredients of a generative CAPP system. In the early eighties, knowledge based CAPP made its introduction using AI techniques.

The generative CAPP systems are often called expert CAPP systems because these systems utilise the expert process planning methodology.

SYSTEM APPROACHES IN CAPP

Developments in computer based planning are attempts to free the human from the planning process and to eliminate decisions required during design and planning. CAPP has various levels of human intervention according to used approaches. The two principles are principally different.

In the first principle - Group Technology based approach or so called variant approach - a planner retrieves the plan for similar components using coding and classifications of parts. The planner edits the retrieved plan to create a variant to suit the specific requirements of the component being planned. This technique is based on the principle that geometrical and technological similar parts have similar process plans. The computer aid is used to assist in identifying similar plans, retrieving them and editing the plans according to the geometrical difference.

In variant CAPP system the process plan is assigned for the whole part according to the global part information. Parametric information between the technological operations and the part feature does not exist.

In the second approach – generative method - the process plan for a new part is automatically synthesized. The generative CAPP system, also called exact system, creates the process plan from information available in manufacturing databases according to a CAPP methodology. The CAPP system operates without or with little human intervention. Each of the part features can be manufacturable by several technological operations. For individual feature from manufacturing knowledge base, the technological operations (technological transformer) ensure the required part properties are generated. From a set of convenient technological transformers an optimal aggregate of technological transformers is extracted. There is parametric information between the technological operation and the part feature.

The manufacturing knowledge is one of the basic information bases for automated process planning. The manufacturing knowledge in the variant CAPP systems is placed in standard plans for each family group. Knowledge is complexly expressed in manufacturing, fixturing and heat treatment instructions. The knowledge is represented in textual or coded form and is not systematicly divided into knowledge bases.

The knowledge in the generative CAPP systems is placed in the individual bases. They should consist of information on manufacturing methods, manufacturing equipment, fixturing, heat treatment, product feature structure, etc. The individual bases are mutuall in relation. The knowledge is directly expressed and is represented in various representation schemes (production rules, frames, decision trees, decision tables, semantic nets). The two basic approaches require the different describing of the part properties. The generative CAPP system needs the unambiguous description of the geometrical, topological and technological part properties. For variant CAPP approach it is convenient to have ambiguous part information, for example some of well-know GT codes (Opitz, CODE, Miclass, Dclass code).

5

Modern Strategic Roles of Manufacturing Technology

INTRODUCTION

Currently, manufacturing is viewed as a simple process of transforming materials into products mostly. Trying to propose ideas to make manufacturing work more efficiently and/or effectively, most studies take their outset in offering customers what they want at the lowest possible cost. However, this view no longer suffices as the environment of manufacturing has faced significant changes in the past decade. In fact, the most notable challenges for manufacturing are increased levels of complexity and uncertainty coming from increased globalization, of markets and operations, the diversified demands of customers, drastic reductions in product lifecycles, and manufacturing and ICT technology progress. In a word, the knowledge base for manufacturing has become more complex and this process is likely to continue.

Therefore, it is quite important to change our perspectives on manufacturing, from a resource-based to knowledge-based view; from linearity to complexity; from individual to system competition; and from mono-disciplinarity to trans-disciplinarity.

Manufacturing strategy is not just about aligning operations to current competitive priorities but also about selecting and creating the operating capabilities a company will need in the future. When manufacturing starts to play a somewhat different role, as sketched above, this opens for a discussion of current thinking and practices approaching manufacturing from the (traditional) best fit, focus and trade-off perspectives.

Thus, this paper will focus on the changes in the strategic roles of manufacturing that are initiated by the challenges mentioned above. It begins with a brief review of the literature on strategic roles of manufacturing. By pointing out shortcomings in existing research, the main questions of this paper are formulated, and the research method employed to research these questions described. Then, four detailed cases are introduced and analyzed to provide the basis of four new strategic roles discussed next. The paper is concluded with a summary of the findings and directions for future research.

LITERATURE REVIEW

The notion of manufacturing strategy as a separate but related functional component of a business unit strategy was first put forward by Skinner in his two papers. Currently, the dominant view is that research on manufacturing strategy consist of two categories—content research and process research. According to Adam and Swamidass (1992), content research addresses the decision scope of manufacturing strategy, which includes two core elements. The first element is a statement of "what the manufacturing function must accomplish", or the "manufacturing task" (Skinner, 1978), which refers to critical competitive capabilities, e.g. quality, cost/efficiency, delivery/ responsiveness, flexibility, innovation and customer service.

The second element of a manufacturing strategy is defined by the pattern of manufacturing choices that a company makes, namely structural or "bricks and mortar" decisions about facilities, technology, vertical integration, and capacity and major decisions about the manufacturing infrastructure, such as organization, quality management, workforce policies, and information systems architecture.

In process studies of manufacturing strategy, there are also two mainstream theories, manufacturing strategy as a top-down (directed, intended) and as a bottom-up (emergent) process, respectively.

In the literature about strategic roles of manufacturing, the role of manufacturing is "defined" as the strategic contribution of manufacturing to the competitive strength of a company. Hill (1983) proposes various concepts and ideas and gives practical examples of how to develop the strategic role of manufacturing. Many authors in this line of research regard the positioning of manufacturing in its wider environment as a question of fit and focus. Hill (1985) defines the manufacturing task in terms of the capabilities that are critical to meeting customer demands. This means that manufacturing plays two key roles: qualifying for, and winning orders in, the market place. Within the perspective of fit and focus the strategic role of manufacturing can also be described by its location and contribution to the value chain of a company, following the ideas of Porter (1985).

Based on empirical findings, Wheelwright & Hayes (1985) identified four different roles (stages) of manufacturing: internally neutral, externally neutral, internally supportive, and externally supportive, which they saw as a maturity model of strategic manufacturing, proposing that manufacturing companies make a choice as to how they compete (Child, 1972). Gilgeous (2001) provides some evidence for the characteristics of strategic manufacturing effectiveness, to provide an empirical validation of the strategic role of manufacturing and to make the structure of the four-stage framework explicit.

Voss (1995) introduced three paradigms of manufacturing strategy, respectively: competing through manufacturing; strategic choices in manufacturing; and best practice. In the first paradigm Voss included order winners, key success factors, capability, generic manufacturing strategies and shared vision. In the second paradigm he included contingency approaches, internal and external consistency, choice of process, process and infrastructure and focus. In the best practice paradigm, Voss included world-class manufacturing, benchmarking, process re-engineering, TQM, learning from the Japanese and continuous improvement. In his 2006 revisited the paradigms and stressed that there is a need for adding more dimensions to the strategic role of manufacturing, following the increased distribution of manufacturing and increased complexity (Voss 2006).

Most researchers regard manufacturing "simply" as a process of transforming materials into products and propose ideas to make manufacturing work more efficiently and/or effectively. Manufacturing strategy, then, concerns the question of how to pursue specific competitive priorities efficiently and effectively according to changes in corporate strategy and the internal and external environment. In the focus and fit perspective dominating this approach, the emphasis is on offering customers what they want.

However, it is less clear how much freedom manufacturing should have to develop competences that go beyond immediate requirements, but ever more authors advocate the idea that manufacturing competencies and their development may also create competitive advantage for the company. Then, manufacturing strategy is not just about aligning operations to current competitive priorities but also about selecting and creating the operating capabilities a company will need in the future. In effect, the role of the manufacturing function starts to change. Rather than simply carrying out their assigned mission, they also have the authority to redefine that mission. This opens for a discussion of current thinking and practices of manufacturing related to the traditional best fit, trade-off and role perceptions. Moreover, it may change our paradigm of manufacturing based on physical resources to manufacturing based on knowledge.

Johansen & Riis (2005) propose another way of characterizing the strategic role of manufacturing based on the thesis that an industrial company can occupy a number of different positions in the supply chain. In view of the close interaction between the various functions of an industrial company it is difficult to identify a strategic role that manufacturing plays alone. Based on a survey including approximately 1,800 Danish companies, they identify five different roles. Full scale production is carried out exclusively by manufacturing, whereas the following four roles are supporting one or more functions, such as ramp-up (sales and product development), prototype production (product development, sales and sourcing), benchmarking (sourcing), and laboratory production (product development). However, these authors do not account for how thy arrived at these five roles, nor do they provide empirical support for their findings or analyze the five strategic roles in detail.

Following the above discussions, the objective of this paper is to replicate, and elaborate on, the work of Johansen and Riis (2005), discuss and, possibly, modify and/or add to the strategic roles for manufacturing these authors identified, and provide more detailed insight into the (modified, new) roles.

METHODOLOGY

Our objective calls for explorative in-depth research, for which, at this stage of theory development, case studies are the most suitable methodology (Yin, 1994). In the next section, four case stories from four different industries are introduced. Open interviews and document study were the main methods used to perform the case studies.

CASE STORIES

Company A is an OEM supplier of medical textiles, which has few but very important customers. Due to a strong price competition in the market for incontinence products, Company A has recently come to recognize the importance of customers' demands. Manufacturing is not longer viewed as the dominant activity, but as a means for realizing customers' needs and obtaining better customer satisfaction.

This change of manufacturing role has called for a greater understanding of which needs manufacturing should fulfill, as well as how these needs are satisfied for Company A. The result of these considerations is that most high-cost manufacturing was moved to factories in Ireland, Slovakia and US, where financial advantages can be picked up. However, on the other hand, central, that is, knowledge and competence intensive manufacturing tasks remained at company' headquarters in Denmark. The starting point is to combine manufacturing competence with product development, so that prototype production and process development are handled at headquarters, where there are two different manufacturing halls.

One manufacturing hall is reserved for the R&D department, and has two primary functions: prototype production and laboratory. Here product developers have privileges to test new ideas and to produce and improve prototypes. They enjoy enough freedom to experiment with new products and new technologies.

This freedom combines with a wish of testing and experimenting, which makes products of Company A so attractive, and thus, its market position so strong. The other production hall handles the running-in of prototypes from the first hall. Here pilot series are made by new products, and are documented with the help of process flow and work-instructions.

Company B focuses on developing and manufacturing unique, customer-specific components and total solutions in the area of plastic and metal technologies. Being an OEM supplier and facing strong competition from factories in China, it is under constant pressures to renew its product portfolio and manufacturing procedures, and price pressure increased, too. Company B used to be competitive, on flexibility and change-over ability, not on price, and needed to develop the capability to combine rapid adaptation to changing demands of customers with the efficiency of mass production, so as to provide specific, high quality and low price solutions to customers. The actually strength of its current manufacturing system is the combination of ramp-up manufacturing and mass manufacturing, that is, the combination of flexibility and efficiency. Although ramp-up manufacturing brings complexity to the production system, it also makes it possible for Company B to maintain manufacturing in Denmark and offer low prices simultaneously. To some extent, manufacturing of Company B could be viewed as a textbook example, which points out that manufacturing in Denmark could also be competitive, as customers demand not only "cheap" products.

Company C is the one of largest kitchen companies in Scandinavia. The operational objective of the case company is to deliver a large range of products to customers in order to satisfy their special demands. There are some clear demands for production, including low cost, high delivery reliability and constantly high quality, combined with the flexibility to produce and deliver kitchens with different configurations and made from different types of wood. Company C has well-developed, mature products and production processes, from component manufacturing to assembly of whole kitchens. Its production system is made up by three departments: component department, special department and assembly department. As their names point out, the component department produces standard components, the special department produces customer-specific components and the assembly department integrates components to make up the whole kitchens. The component department produces according to forecast, while the other two departments produce to customer order.

Company D is one of four SBUs of a big energy company, which merged with another company to create the largest manufacturer of windmills in the world in 2004, delivering approx. 4000 windmills per year. Mainly due to political and logistic issues, Company D follows a strategy that it only holds 10%-12% of its manufacturing in Denmark while the rest is outsourced to local suppliers. Its, consequently small scale, manufacturing system involves all the equipment and processes needed to produce windmills. This in-house manufacturing, which could be viewed the mini version of the manufacturing operations of Company D's partners, acts as a benchmark for those partners. Company D selects proper suppliers and then helps them to improve their performance. In order to support the knowledge transfer to its suppliers and help them improving their performance, Company D mainly relies on documents and a "supervisor corps". Documents can be used as manufacturing the towers is not considered a core-competency and can be classified as low-tech manufacturing. Moreover, all the operations related to tower manufacturing are standard. Thus, it is possible for the suppliers to produce according to standard operating procedures. However, still, different kinds of problems may occur during the various manufacturing phases. To tackle that, Company D utilizes the "supervisor corps". The corps consists of experienced craftsmen (e.g. welders and CNC-operators) who visit the manufacturing sites and assist the external suppliers based on their expertise from the benchmarking manufacturing in Denmark. Thus, the supervisors are responsible for solving problems faced by the external suppliers, while they also bring back manufacturing knowledge to the Engineering and Manufacturing departments in Denmark from their problem solving experiences.

FINDINGS

Table. Characteristics of four strategic roles of manufacturing

Cases	Objectives	Competitive priorities	Key resources	Inter-relationships with other functions
Company A	Developing and testing new processes, new products and production equipment, even new technical/administrative systems	Innovation	Technological resources	Mainly R&D and marketing, while external research centers, universities and customers could be involved
Company B	Establishing a production system to keep pace with increasing demands from technologies or markets as quickly as possible	Flexibility and delivery	Proximity to R&D centers or markets	Mainly between R&D and manufacturing and between marketing/ sales and manufacturing
Company C	Taking part in the company's continuous development and profit making, and being able to live up to quality, price and on-time delivery	Quality, price and on-time delivery	Access to low cost production input factors (sometimes proximity to markets)	Mainly focusing on manufacturing itself, but also working, more or less, as a caller to get help from other functions
Company D	Getting information of a certain production flow, which could be used as a benchmark, to help making some strategic decisions	Flexibility	Technological resources	Mainly between manufacturing and outsourcing or procurement

We summarize the case stories with respect to their strategic roles of manufacturing from four aspects: (1) the objectives, (2) the competitive priorities pursued (quality, cost/efficiency, delivery/ responsiveness, flexibility and innovation), (3) key resources (access to low cost production input factors; proximity to market; use of local technological resources) according to Ferdows (1997), and (4) inter-relationships with other functions.

DISCUSSION

New, different roles for manufacturing

Viewed from a material flow perspective, manufacturing is the last function before products come out and are delivered to the market place. Before manufacturing, different sorts of information, knowledge and materials from different functions come together, and during manufacturing, they are transferred into specifications of manufacturing processes and used to support finished goods production. After completion of the actual production process, manufacturing could be viewed as the starting point of the delivery process. To some extent, manufacturing is arguably at the center of the entire operations of industrial companies, as a "processor" that collects all sorts of information, knowledge and materials from different functions, processes them and transfers them in the form of final products to the market place.

Besides its traditional role, the possibility exists for manufacturing to play additional roles through interactive support, which means manufacturing cooperates with specific functions, serves specific objectives and gives adequate support to these activities, as shown in Company A, B, and D. Generally, because manufacturing is viewed as the center of the entire operations of industrial companies, it is natural that there could be three types of interactive support, namely backward interactive support (upstream), forward interactive support (downstream) and lateral interactive support. Backward interactive support means that manufacturing takes part in activities, e.g. innovation and product development. Many studies refer to this area as, for example, integrated product development or concurrent engineering. Forward interactive support means that manufacturing takes part in activities after completion of the (physical) product, including distribution and after-sales service.

Four new strategic roles of manufacturing

Following this line of thinking, the cases suggest four different strategic roles for the manufacturing function:

- Innovation manufacturing, which takes part in R&D activities and works with R&D (maybe also with marketing) to realize innovations. In this role, the manufacturing function is home to the development and test of new technologies, products and/or management systems (Company A).

- Ramp-up manufacturing, which mainly works with marketing (maybe also with R&D). It aims to establish a production system capable to keep pace with increasing demands for a new product or in a new market. At the same time, it also embraces possibilities of a temporary set-up for the establishment and running-in of an assembly system based on new technologies (e.g. Company B).
- Primary manufacturing, the traditional role, which aims to produce as efficiently and effectively as possible. Manufacturing capacity keeps pace with the demand for the company's products. Pursuing priorities as to quality, price and on-time delivery is one of the most important tasks in this role. Most OM studies take this role as their starting point (e.g. Company C).
- Service manufacturing, which could be viewed as the aggregate of many different possible roles that manufacturing could play, and aims to serve specific context dependent objectives. Benchmarking manufacturing can be classified into this role (e.g. Company D).

The first two roles – innovation and ramp-up manufacturing – involve backward interactive support as they take part in some activities before traditional manufacturing. With regard to the third role, it just concentrates on manufacturing itself and does not provide forward or backward interactive-support. Benchmarking manufacturing provides lateral interactive support, to parallel systems essentially producing the same components or products, but may also interact with upstream and downstream processes. We suspect benchmarking is just one form of many actually providing services to other functions or production systems, inside or outside the company – hence the term service manufacturing.

Our observations are limited to four cases, in which we did not see any forward interactive support. However, it does not mean that this mode is not important. Furthermore, there may still be other strategic roles – further research is needed to investigate this. The four strategic roles of manufacturing differ from the classification of Johansen and Riis (2005) as follows:

- In practice, laboratory production and prototype production are always interdependent and they have the same objective—product development. Moreover, according to our observations, they are in the same place in most of situations. Therefore, we combine them into a new role—innovation manufacturing.
- Actually, comparing with other roles, full-scale production plays a traditional role, which focuses on the primary objective of manufacturing – producing more efficiently and effectively and it is the only role carried out exclusively by manufacturing while other roles normally cooperate with other functions to realize supportive objectives. In order to describe these comparisons more clearly, we use primary manufacturing to replace full-scale production.

- We extend the conception of benchmark production to service manufacturing. Benchmarking production is one role, in which manufacturing services for the sourcing (purchasing) function to get relevant information about suppliers and also provides services to those suppliers. But manufacturing may still have possibilities to service for other objectives. That is why we propose the label service manufacturing as the aggregate of possible roles that manufacturing could play.

We saw no need to modify the ramp-up role proposed by Johansen and Riis (2005).

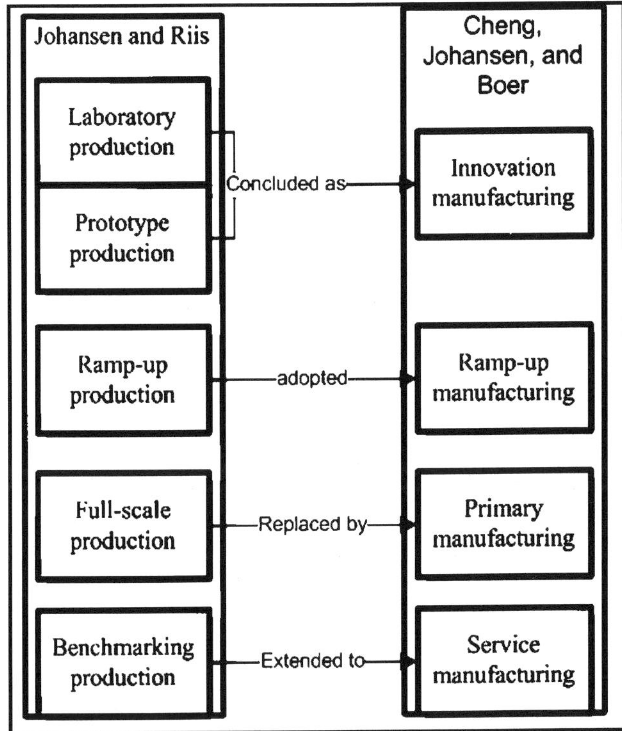

Fig. Comparison of two typologies of strategic roles of manufacturing

Furthermore, it is obvious that the four strategic roles are not necessarily entirely separated from each other. On the contrary, they can be related along two dimensions: time and place.

Firstly, viewed from the time dimension, the four strategic roles seem to represent the typical product life cycle. Innovation manufacturing is relevant during the development phase of new products. Ramp-up manufacturing is used during the introduction and growth phase. Certainly, primary manufacturing satisfies the maturity phase. Service manufacturing could be useful in many phases, depending on service provided and the internal or external customer(s) of that service. Secondly, along with the trend of globalization, it is more and more normal for industrial companies to have several plants spread geographically. And to some extent, the strategic role of plants could be decided by manufacturing roles.

Individual plants could be characterized by the manufacturing role(s) they play and they could be viewed as locating specific manufacturing roles in specific places and giving specific forms of support, depending on their role. This suggestion provides a starting point for discussing the role of plants in networks.

Viewing manufacturing as a support activity

Primary manufacturing is the only role that could be viewed as the traditional one, which mainly focuses on manufacturing itself to support producing efficiently and effectively. Without denying the significance of primary manufacturing, it seems that the contributions that manufacturing makes to competitive advantage in cooperation with other functions will become more important in the future (Riis et al., 2007). According to Porter (1985), manufacturing has always been treated as one of the primary activities in industrial value chains.

But according to the analysis presented above, this point may need to be revised. On the one hand, there is still room for manufacturing to play its traditional primary role. On the other hand, however, the other three roles seem to be better characterized as support functions. Innovation manufacturing supports the development and test of new processes, new products and production equipment, even new technical/administrative systems; ramp-up manufacturing supports the introduction and growing demand for new products and/or from new markets; and benchmarking manufacturing parallels production systems and informs the (out)sourcing function.

With the competitive environment changing continuously and ever faster, it is predictable that the indirect strategic roles of manufacturing (innovation, ramp-up, and service manufacturing) will become increasingly important in the future, which points to the need to focus attention on developing competencies in managing the interplay between manufacturing and other functions, such as sales, product development, sourcing, distribution and after-sales service.

As the indirect strategic roles come into focus and operations take place globally, new competencies are called for. Traditionally, emphasis has been placed on knowledge and know-how about production processes; and this represents an important challenge for key processes. But increasingly the capability to manage the complex interplay between many actors involved in a value chain will become important.

The strategic roles of manufacturing imply companies need to develop competencies in this area (Riis et al., 2007), but it is unclear as yet exactly what these competencies involve.

CONCLUSION AND FUTURE RESERACH

This paper argues that manufacturing should no longer be viewed as a simple process of transforming materials into products as efficiently and effectively as possible. Viewed as the center of the entire operations of industrial companies, the possibility emerges for manufacturing to play different and equally important role.

Through backward, forward and lateral interactive support, manufacturing could take part in other relevant activities before or after manufacturing and support operations of other (parallel) functions. Thus, besides traditional role – primary manufacturing, this paper introduces three other strategic roles: innovation manufacturing, ramp-up manufacturing and service manufacturing.

It is predictable that the discussion of these new strategic roles of manufacturing will become more and more important. Firstly, the globalization means that it is more and more normal for companies to have new product development in one place, ramp-up manufacturing elsewhere, and full capacity production in a third location. Obviously, in different phases, manufacturing could have different effects, that is, play different roles. Secondly, in different situations, manufacturing needs input from or provides input to and, thus, has to cooperate with different functions to realize its own role.

In order to know more about how manufacturing works with different functions in different phases and how manufacturing acts as a platform to support different activities, the four strategic roles of manufacturing are valuable to be researched. Thirdly, following the cooperation with different functions, the knowledge used and created may differ from phase to phase. In order to make this knowledge aspect clearer and easier to be researched, it is, again, valuable to focus on four roles of manufacturing.

This paper tries to view manufacturing from a new angle, going beyond fit, focus and trade-offs. The findings provide the basis for future research, mainly focusing on three parts. Firstly, the paper is based on four case studies. Further research is needed to refine and, possibly extend the findings. In practice more roles may be found. And, then, our understanding of each of the roles is rather limited.

Second, from a life cycle perspective, it is necessary to focus on knowledge transfer between the different strategic roles. Finally, globalization leads to, amongst others, the development of dispersed plant and supply networks. The interaction between the strategic roles in such networks is a third important area for further research.

GLOBAL MANUFACTURING STRATEGY

As global competitiveness intensifies, companies get new products into all major markets quickly and respond distinctively to these market needs.

The dramatic advances in manufacturing technologies and organizational infrastructures all over the world demands a right manufacturing strategy that provides competency in their operations (Roth et.al. 1989). Moreover, the global economy that is rapidly evolving into an integrated system presents both opportunities and threats to global manufacturers. These new frontiers warrant the analysis of global manufacturing strategic issues for competitive survival.

The critical problem of the global manufacturers is to initiate and operationalize a competitive strategy that would evolve the firm to meet the challenges in the global market (Scully 1993). Manufacturing being highly technology dependent, its implementation of strategy requires a strong foundation of technological resources and technological capabilities. The accumulation of resources and constant up-gradation of the capabilities would provide competitive strengths for manufacturers in global scale. But studies in this arena hardly linked strategies with the resources and capabilities of firms in an articulated way.

STRATEGIC ISSUES IN GLOBAL MANUFACTURING

Manufacturing is generally viewed as transformation process in a narrower perspective. "Install the latest technology" is the recurring theme among the manufacturers to attain competitive leverage. There are divergences in the emphasis on manufacturing strategies and its practices in different economic systems (Ettlie 1996). But the proper approach is to embrace the critical strategic issues and translate them into actions at all levels of management in a broader perspective (Leong and Ward 1995).

Essentially, manufacturing strategy comprises a set of well-coordinated objectives and action programs aimed at securing sustainable advantages over competitors. It deals with the importance of manufacturing in business strategy and guides manufacturing decisions at the functional level of the company.

The global manufacturing strategy involves manufacturing operations spread among many countries. Leong and Ward (1995) suggested six Ps of manufacturing strategy that includes planning, proactiveness, pattern of actions, portfolio of manufacturing capabilities, programs for improvement, and performance measurement. These six Ps individually provides a distinct view, which is partially revealing about the strategic intentions and capabilities of global manufacturing. Roth (1989) emphasized on priorities of the set of action programs including flexibility, quality, delivery and price as global manufacturing strategies.

Hayes and Wheelwright (1994) outlined quality, production planning, technology, and the work force in explaining global manufacturing strategy. They catego-rized four stages of manufacturing strategy distinguishing orientations from global to general such as, internally neutral (production simply makes the product and ships it), externally neutral (manufacturing merely meets the standards set by competition), internally supportive (manufacturing attempts to become unique compared to its competitors), and externally supportive (manufact-uring pursues uniqueness on a global scale, and becomes a world-class competitors). These four stages range from reactive (internally neutral or defensive strategy) to proactive (externally supportive or offensive strategy).

The fourth stage illustrates the global manufacturing strategy. Production-allocation approach is a tool for enterprises in planning their global manufacturing strategy (Vos 1991).

The prime objective of this approach is to lower the transportation and manufacturing costs in the long run considering locations of the plants and the extent of vertical integration (forward and backward). Sweeney (1991) identified four types of generic manufacturing strategies such as, marketeer, caretaker, reorganizer and innovator which progressively increases its proactive intensity and leads to global manufacturing strategy (innovator strategy).

RESOURCE AND CAPABILITY-BASED GLOBAL MANUFACTURING STRATEGY

A manufacturing strategy is defined as a statement of how manufacturing supports the overall business objectives through the appropriate design and utilization of manufacturing resources and capabilities (Joseph 1999). In the process of formulating it could be explained from matching dimension where strengths and weaknesses of the company are matched with opportunities and threats (Grant1991). Resources and capabilities provide a basis of intensity of strengths or competencies to global manufacturers to avail the opportunities and to overcome threats.

Technological Resource-Based Strategy

Technological resources enable manufacturers to generate above-normal rates of profit and sustainable competitive advantage (Mata et. al. 1995; Grant 1991; Dierickx and Cool 1989; Oliver 1997). Its value is determined in the interplay with scarcity, demand and appropriability and is traded in open market situation (Collis et. al. 1995).

Few researchers have classified technological resources in strategic perspective. In manufacturing organizations it is classified into technoware, humanware, orgaware, and inforware.

In this classification, technoware refers to the tangible and palpable part of the machineries, humanware refers to human skills needed to realize the potential of technoware, orgaware refers to the support net of principles, practices and arrangements that govern the effective use of technoware by the humanware, and inforware refers to accumulated knowledge needed to realize the full potential of the technoware, humanware, and orgaware. It is important to note that technological resources are function specific and as such, all the components are required to be present for manufacturing processes (Saha and Islam 1998).

Table. Strategic Issues in Global Manufacturing

Proponents	**Strategic issues**
Contractor and Loarange (1988), Nassimbeni (1998), Wildeman, L. (1998)	Global alliances and networks in global manufacturing include supply relationships, agreements and joint ventures, and regional industrial systems.
Roth et. al. (1989), Ferdows et. al. (1985), Young et.al (1992)	Global manufacturing strategies are characterized by the sets of strategic action programs that link with competitive priorities of flexibility, quality, delivery and price.
Toni et. al. (1992)	Integration and co-ordination of all activities in the value chain
Scully and Fawcett (1993)	Explores the vital linkage between global manufacturing and strategic advantage of logistics and production costs.
Hayes and Wheelwright (1994), Harrison (1998)	Four stages of manufacturing strategy ranging from reactive (internally neutral) to proactive (externally supportive). They are (i) Internally neutral. Production simply makes the product and ships it. (ii) Externally neutral. Manufacturing merely meets the standards set by competition. (iii) Internally supportive. Manufacturing attempts to become unique compared to its competitors. (iv) Externally supportive. Manufacturing pursues uniqueness on a global scale and becomes world-class competitors.
Leong and Ward (1995)	Global manufacturing strategy should be viewed from a broader perspective that includes six Ps: planning, proactiveness, pattern of actions, portfolio of manufacturing capabilities, programs for improvement, and performance measurement. Each P is offered as a distinct view, which is partially revealing about the strategic intentions and capabilities of manufacturing.
Vos (1991)	Emphasized on the locations of the plants and vertical integration (forward and backward) to reduce transportation and manufacturing costs in the long run.

Table. Technological Resources - Examples

Technological Resources	**Examples**
Technoware	Tools and equipment (manual and powered), Machineries (general purpose or special purpose), Vehicles, Other facilities (automated, numerically controlled, computerized).

Humanware	Skills (ability to comprehend and use job related components, ability to mobilize, setup and utilize technology components for work, ability to optimize use of available technology components for all tasks), craftsmanship, expertise, dexterity, creativity (ability to undertake component innovation activities for better performance).
Orgaware	Techniques and methods (tradition-based work organization, education and experience-based work facilitation, systems analysis and optimization, reengineering and innovation) organizational networks, management practices.
Inforware	Facts and formulae (documented knowledge for acquisition and optimal performance), design parameters, specifications, manuals (operations and maintenance), theories (state-of-the-art knowledge for innovation).

TECHNOLOGICAL CAPABILITY-BASED STRATEGY

Technological capabilities explain why firms are different, how they change over time and whether or not they are capable of remaining competitive (Patel and Pavitt 1997). It is the ability or skill of the firm at coordinating its resources and putting them to productive use that leads to achieving competency. Technological capabilities are defined and classified by number of researchers. The differences in opinions are observed among them. Since early 90's, emphasis has been put mainly on acquisitive, operative, innovative and supportive technological capabilities. The capabilities represent various functions of a firm involved in manufacturing. The functions are in turn associated with manufacturing strategies. Hence, technological capabilities and manufacturing strategies are closely interrelated. The descriptions of four technological capabilities in brief are presented below:

- Operative capabilities - Ability for operating and controlling plant and equipment, planning and controlling production activities, providing information support and networking for operations, maintaining the plant and equipment in good order.
- Acquisitive Capabilities - Ability for carrying out detail engineering study, independently searching for good technology sources, assessing technologies offered, deciding technology transfer mode, and negotiating terms of technology transfer.
- Innovative capabilities - Ability for duplicating acquired technology, adopting and carrying out improvements in imported technology, carrying out own technology development plan.
- Supportive Capabilities - Ability for undertaking project planning and execution, obtaining funds for prototype development and

modernization, planning and implementing human resource development, identifying and developing new markets for the firm's existing and new products.

Global Manufacturing Strategic Issues with Technological Capabilities

The prime basis of formulating global manufacturing strategy is technological capability. It has been focused differently in various researches and thus lacks an integrative understanding to the manufacturers. As such, strategic issues in global manufacturing are translated into functions that include acquisitive, operative, innovative and supportive activities. Through performing these activities it is convenient to implement manufacturing strategies at the global marketplace. The linkages between global manufacturing strategic issues with technological capabilities are presented in Table. The strategic issues in global manufacturing are shown in left side of the table and the right depicts different functions performed by manufacturers. Integration and coordination of activities in value chain on global scale is performed through primary and supportive activities that are in practice acquisitive, operative, innovative and supportive in nature. Likewise, competitive priorities could be achieved by emphasizing on cost, delivery, quality, and flexibility. To achieve the performance four types of functions are essential for all global manufacturers.

A Strategic Model for Resource and Capability-Based Global Manufacturing

The core of the global manufacturing strategy is to achieve manufacturing competency in its operations. Resources and capabilities provide a basis to shape manufacturing strategies that eventually builds manufacturing competency. The interrelationships of global manufac-turing competency with technological resources, technological capabilities, backward supports, and forward supports are shown in Figure. It comprises mainly of three parts. The main part depicts the global manufacturing competency influenced by technological resources, technological capabilities and supports. Each technological capability is derived from use of all the technological resources. Technological resources and capabilities along with the backward and forward supports result manufacturing competency. The second part deals with the requirement of resources and capabilities for manufacturing and the last part focuses upon forward and backward manufacturing supports in terms of suppliers and customers. The components of the model and their interrelationship will result manufacturing competence at global scale.

TECHNOLOGICAL INNOVATION AND ROLE OF TECHNOLOGY STRATEGY

The current dynamic environment demands all organizations to change

– both radically and incrementally. Sustainable development cannot happen without innovation. It is very essential for an organization to change the way it operates and also change the products and services it provides. It calls for an organization that encourages experimentation, constantly monitors the environment, evaluates its own performance and is committed to continuously improve performance. Managing innovation is multifaceted and involves various aspects.

A stream of research linking innovation to the knowledge base of the organization highlights the importance of various critical factors which enable the acquisition and application of knowledge to promote innovation. It mentions that implicit knowledge can be managed indirectly by managing various factors which contribute to an organization's culture, structure, technology and leadership. According to this stream of research innovation can be promoted by managing the tacit and explicit knowledge embedded in these factors.

One of these factors, technologies is continuously changing and is a critical contributor to the turbulent markets. Moreover, technology as an intangible knowledge based asset increases the potential of innovation management. Ettlie & Bridges (1983) suggest that technology policy reflects the innovative attitude of an organization and its commitment to innovation. While technology can play an important role in an organization's innovation process, too much technology might even stifle innovation by standardizing the existing workflows and processes. The long run competitive position of an organization is therefore dependent on how prudently organizations manage their technological asset bases. Firms differ because they develop competencies in different technologies. Asymmetries in knowledge endowments may provide a basis for technological heterogeneity. On the other hand firms may also differ in the way they exploit or deploy the available/ acquired technology. (Nesta & Dibaggio, 2003).

Today, the strategic management of technology is considered one of the most important issues in literature (Zahra & Covin 1993). "A company can use technology to create a competitive advantage by creating barriers that deter entry of rivals, introducing novel products or technology processes that attract new customers, or changing the rules of competition in the industry." (as quoted in Zahra, 1996). The creation, development and application of technology can form the basis for the success of firms. Whereas the focus of technology in low technology organizations is primarily to utilize and expand technology innovation, technology in high technology industries is one of the critical factors for determining the future success of firms.

Growth through innovations might not necessarily take place through break-through innovations. Rather it involves innovative ways of integrating existing or developing technologies into high value solutions. These innovative ways span across the entire value chain and require integration across internal and external business boundaries. The technological choices of a firm are usually clarified in its technology strategy i.e., the plan that guides the accumulation and deployment of technological resources and capabilities (Zahra, 1996). According to Kaplan & Norton, "In our practice, however, we observed that no two organizations thought about strategy in the same way."

In spite of the increasing importance of innovation and the role played by technological capabilities in a firm's growth trajectory surprisingly, little is known how technological innovation in different organizations is driven by their technology strategy. Thus research is necessary to determine how technology strategy guides the acquisition, deployment and abandonment of technology to promote innovation.

This paper begins by reviewing some of the key literature in technology strategy and technological innovation. The literature hints at the importance of various organizational factors which promote innovation in the organization. On the basis of the literature review a conceptual framework has been developed to explore the relationship between technology strategy, technological innovation and organizational factors, followed by the outcomes. The government rules and various policies also affect the technology strategy of firms. The paper further discusses the role of technology strategy in technological innovation.

TECHNOLOGICAL INNOVATION

Innovation includes both product/ service and process innovations. Product innovations are products that are perceived to be new by either the producer or the customer; the latter includes both end-users and distributors. Process innovation refers to new processes which either reduce the cost of production or enable the production of new products. Service innovation involves change in the process of delivering existing services or the development of completely new kind of services. (Leiponen, 2005). It can also include the application of advances in technology to an existing product so as to create a new product.

According to Chesbrough (2003), " The innovation process has changed from a somewhat closed, singular innovation project to a continuous active and 'open' innovation system." (Duin, Ortt & Kok, 2007). Researchers have brought out the importance of collaborations and networks for technological innovation. Collaborations across functions, products and divisions and trust in these networks (Rycroft, 2006) enable firms to thrive by combining capabilities across businesses. Innovation occurs only when an organization couples organizational know-how with needs of the users. The cyclical innovation model views innovation process as continuous interaction between developments and changes in markets, product and services, technology, and science. According to Van de Van (1986), innovation, or the newness of an idea can include both technological innovation (new technologies, products, and services) and administrative innovations (new procedures, policies and organizational forms). (Hung, 2004). "Technological innovation pertains to products, services and production process technology; it is related to basic activities and can concern either product or process." We can, therefore, assume that technological innovations (outcomes) are the result of product/ process development activities and market development/ service improvement activities.

"Service and process technological innovations comprise technologically new or significantly improved products and services. An innovation has been implemented if it has been introduced on the market (service innovation) or has been used within a production or delivery process innovation. Usually, the introduction of service and process technological innovations involves a series of scientific, technological, organizational, financial and commercial activities." For a service to be considered as technologically innovative its characteristics and mode of using should either be completely new or should have been significantly improved qualitatively or in terms of their performance and technologies used. A technologically innovative service may involve the use of radically new technologies, a combination of pre-existing technologies or new knowledge. Process innovation, consists of the adoption of a production or a delivery method which is new from a technological point of view. Such adoption may involve changes in equipment, organization of production or a combination of both. The intention of process innovation may be either to produce or deliver innovated services or to improve the production or delivery of existing services which would not have been possible by using the pre-existing production methods.

While some innovations are technology based (personal computers, auto fuel injections etc.), other innovations like new products and services in financial services are facilitated by technology. As adoption activity is a form of innovation, high adoption rates reflect high levels of innovativeness. Technological innovations can seldom be developed by a single firm in a vacuum of an industrial environment. Many complementary innovations are usually required before a particular technology is suitable for commercial application. Many innovation based strategies are based on the unique market application of an existing integrated set of technologies rather than requiring technological break-through.

The technological innovation process consists of four broad stages of Problem Recognition/ Idea Generation, Technology Selection, Solution Development and Implementation. (Narayanan, 2007). There has been a worldwide change in the technological innovation process – both in the way the research is organized and the way the new technology is commercialized. The traditional linear approach – R&D, prototyping, manufacturing startup, marketing and distribution all in-house has given way to less vertically integrated structures, which involve collaboration with other industry participants. Patents are an important instrument for the protection of innovations. Industry surveys have revealed that firms use patents as an appropriability mechanism to meet their 'strategic motives' of use of patents as negotiating levers, as tools for prevention of infringement suits or to block innovations from competitors and also to capture the extra value of innovative efforts. Firms also use the extra market power accumulated through patents to control the diffusion of innovation and results of research. (Sampath, 2007). However, patents come with their own drawback. They provide innovators with longer monopoly periods and tend to increase monopolistic sectors within the economy. This deters both the improvement of existing products and the invention of higher quality goods by outside firms. An innovator can reap the benefits from his innovation either by licensing his patent protected knowledge or by deploying the innovation in products or processes. While in the former case the firm extracts a fee from whom he licenses, in the latter case he surpasses his competitors by increasing the innovation-related sales and/ or profits of his own firm.

The successful commercialization of technological innovation to a large extent is determined by the utilization of complementary assets and knowledge which might be embodied in the firms marketing, manufacturing and after-sales support activities. To the extent the complementary assets are critical or are favorably positioned with the competitors; decide the firm's strategy with respect to the sourcing of complementary assets.

Technological innovation not only serves as an important competitive tool but also plays an important role in improving the firm's performance. Although the relationship between innovation and financial business performance is not simple and well understood, prior research indicates a strong interaction between growth in sales and different innovations; between expenditure on R&D (R&D as a proportion of sales) and new product announcements and with performance measures such as value added and market to book value. A few of the effects which can be associated with service process innovations are: improvement to the service provider's productivity and flexibility; more rapid production and/ or delivery of services; improvement in quality of services provided; increased market presence through a wider range or a more "user-friendly" set of services. Product-process innovation has been found to have a positive impact on financial results (as compared to goals), market position and bargaining power.

TECHNOLOGY STRATEGY

Although, the concept of technology strategy has been a part of management of technology literature since the late 1970s there is still a lot of debate regarding how to define technology strategy. It can be briefly and broadly defined as a portfolio of choices and plans that a firm uses to address the technological threats and opportunities in its external environment.

According to (Meyer, 2008), "The operational expression of a technology strategy is the set of projects that an organization wants to implement. Determining a strategy is selecting the projects and the portfolio of projects." The broad objective of technology strategy is to guide a firm in acquiring, developing and applying technology for competitive advantage.

According to Ford & Thomas (1997), "The firm is characterized not only by the configuration of its own technology, but in addition, by its relationships with and linkages to the systems – or discrete technologies – of others' such that a 'meaningful technology strategy is inevitably a network strategy." (Davenport et.al., 2003). According to Ford and Thomas, 'the key factor in technology strategy is for the firm to understand the relative importance and location of the range of technologies both internal and external that it must employ.' However, for a dynamic technology strategy which is receptive to changes in the environment, it is imperative that the development of technology is embedded with other strategic choices with which a firm co-evolves.

Another important aspect of technology strategy is its implementation. Technologies are not characterized by pre-determined level of efficiency. The level of operational efficiency attained through a technology depends on the efforts made by a firm to assimilate and modify it to suit specific requirements. (Deraniyagala, 2001). One of the key aspects of innovation management is the capacity to develop and implement a sound technology strategy. Technology strategies, however, are not confined to high-technology industries. Capacity-driven or a customer-driven industry too requires a technology strategy. Such strategies determine the choice of technical capabilities and available product and process platforms of the firms. In fact, companies which do not see the need to manage technology as a formal process are dis-advantaged.

Technology strategy of an organization has to be constantly reviewed and monitored so as to cope with the dynamic environment. Feedback loops built into the technology strategy are an important way to adjust and reconfigure the technological capabilities. According to the various dimensions of technology strategy of a firm can be understood through the substance and enactment of technology strategy.

SUBSTANCE OF TECHNOLOGY STRATEGY

Competitive Strategy Stance

Technology strategy as an instrument of more comprehensive corporate and business strategies defines the role which technology can play as a source of competition to sustain advantage in product differentiation or cost or to develop new products and lines of business. Researchers have brought out the importance of fit between technology strategy and other strategies (

business, manufacturing etc.) of a firm in order to gain competitive advantage. A fit between strategies enables an organization to improve its business performance.

Technology leadership can be discussed in terms of timing relative to rivals or the commercial use of new technology. Firms wishing to take on the technology leadership position must make a choice with respect to lead time by which they wish to lead their competitors. According to Lieberman & Montgomery (1988), firms undertaking a pioneering or a technology leadership posture focus their research and development initiatives on creating innovations based on state-of-the art technologies combined with possible first-mover advantages. Companies often use pioneering to capture premium segments, achieve economies of scale, set industry standards or control distribution channels.

Leadership strategies are however, dependent on the efforts of highly-skilled employees who are associated with external networks of specialized workers and expert legal advice. Technology-follower firms, on the other hand put emphasis on strategies that involve low-cost manufacturing of proven products and technologies. A company can outperform rivals only if it can preserve the difference which it has established. The growing popularity of outsourcing reflects the growing recognition that it is difficult to perform all activities as productively as specialists.

Value Chain Stance (Scope)

The scope of technology strategy, with reference to the value chain, determines the set of technological capabilities that the firm decides to develop internally. This set of technologies is called the core technologies and with respect to these the firm needs to decide whether to be a leader or a follower when bringing these technologies to market. The other technologies used by the firm are the peripheral technologies. The scope, in other words, guides a firm to identify the mix of product and process technologies which will comprise the firm's portfolio. Through the value chain stance a firm can decide whether to deploy its resources for a few technological capabilities over which it has competitive advantage or to diversify itself into developing number of technological capabilities. The scale and business focus of a firm decides the scope of its technology strategy.

The Value Chain analysis by Porter examines the operations of the firm in terms of the technologies which are used in each segment and how the different technologies interact with each other. The increased complexity of products and production processes over time has led to diversity in the technology portfolio of many organizations. According to Granstrand (1998), firms with technologically diversified portfolio have larger integration, coordination and communication costs which can be controlled to the extent the technologies are coherent.

Resource Commitment Stance (Depth)

The depth of a technology strategy can be expressed in terms of the technological options available with the firm. The intensity of resource commitment to technology determines the depth of a firm's technology strategy.

Greater technological depth enables an organization to be flexible and respond to new demands of the customers. In general, broader the scope of a firm's technology strategy and greater the firm's commitment to leadership position, greater the resources it will have to commit. Greater the resources committed by an organization to R&D and technology development, greater the organizational slack which an organization can use to meet the changing demands in the marketplace. The technology choice of a firm can also be explained as a trade-off between fixed and variable costs; with higher the investment in fixed costs (advanced technologies) less a firm needs to spend on variable costs. (Yin, 1998).

Management Stance (Organizational Fit)

This refers to the extent an organization can structure itself so as to meet the organizational requirements flowing from competitive, value chain and resource commitment stances. According to Steele (1975), decentralized laboratories are more suited for application of technologies and not for generation of new technologies.

However, decentralized laboratories due to close coupling with markets and other corporate functions are better suited for products related research. (Malecki, 1980). Divisional R&D or the M-form organization structure where technological innovation is decentralized in the various divisions is primarily concerned with applied R&D to address the specific requirements of on-going product development and less concerned with developing fundamental R&D required for building distinctive technological capabilities. (Christenson, 2002).

ENACTMENT OF TECHNOLOGY STRATEGY

According to Christenson and Burgelman, how an organization enacts it technology strategy reflects the substance of technology strategy. Technology strategy is enacted through the appropriation and deployment of technology.

Sourcing

The field of technology sourcing is important for innovation in firms as it determines where and how technological knowledge is to be used to develop new products or processes is acquired. (Hemmert, 2008). Internal sourcing of technology depends on the firm's R&D capability. One of the major purposes of the R & D spending is to influence the future investment favorably, either by lowering costs or by increasing returns. In organizations, R&D is supported as an overhead expense and is a critical requirement for exploratory research efforts and for developing or maintaining technical expertise in areas which are judged to be essential for future competitive advantage. R & D is usually measured in terms of input (R&D to sales ratio) or seen as a company resource (does the company in question have R&D facilities or not). (Harmsen, Grunert, & Declerck, 2000). A firm's R&D efforts not only contribute to generation of new knowledge but also to building its absorptive capacity. (Cohen & Levinthal, 1990). Although a risky option, because the success of the technology is not guaranteed in the market, internal development of technology results in proprietary technology which a firm can use to create an advantage.

A global R&D strategy however requires the development of in-house technical and organizational learning competencies in complex alliances with other firms and international centers of research excellence. Subsequent commercialization depends on a string of global inter-firm alliances in complementary technologies and the pooling of financial, production, and commercialization costs and risks. (Kaounides, 1999).

External sourcing of technology can take place through licensing, strategic technology agreements, mergers and acquisitions or through recruitment of staff with appropriate knowledge. A very relevant point with reference to external sourcing of technology is that, since the technologies available in the market are also available to the competing firms, they cannot offer a sustainable competitive advantage. A mixed internal/ external approach may involve funding of joint ventures, collaborative research or strategic alliances. It stresses on the complementarity between in-house R&D and external know-how. According to (Mowery & Rosenbloom, 1989), " cooperative research programs alone are insufficient ...more is needed, specifically the development of sufficient expertise within these firms to utilize the results of externally performed research." (as cited in Veugelers, 1997,p.304).

However, decision to buy technology requires a much wider knowledge base than being simply aware from where a technology can be bought cheaply. Competitive advantage is gained only from cluster of technologies which are highly synergistic. According to Teece (1977), it not only involves identifying the right technology for a particular purpose but also an understanding of the means by which it will be integrated with the firm's traditional technology. The decision to source technology internally or externally is however dependent on two critical factors- the ease of appropriability (extent to which benefit from research activity can be captured) and the transaction costs (the ease with which contracts for the purchase or sale of technology can be written, executed and enforced without leading to unexpected outcomes that impose large costs on one or both parties. The extent to which a particular technology is protected by intellectual property rights also effects the decision of a firm to source internally or externally. Although endogenous technology strategy which lays emphasis on internal R&D efforts of an organization enables an organization to develop a strong technological base which is inimitable (Howellis, 1997; Yin, 1998) the importance of complementing internal technology with external technology cannot be denied. The opportunities for appropriating the benefits of internal R&D also influence a firm's commitment of resources to the process. ((Zhang & Liu, 2007)

Product and Process Development

According to, "Product development can be described as the process that identifies a market opportunity and transforms it into a product available for sales." Technology strategy is also enacted by deploying technology to develop products. While taking a decision to develop products firms need to take a decision whether technology would drive the development of the product (technology push) or product development and/ or market development would drive the development of technology (market pull). Every product is composed of a number of technologies. Firms infuse technology into new products either by bundling or by disruptive technologies. Through bundling, firms combine the different elements of their product lines into bundles. Radical innovations, and sometimes modular or architectural innovations, deploy disruptive technologies.

The benefits of information technologies like desktop software and web-based tools for different stages of the New Product Development (NPD) process are being widely recognized. However, it is essential for technology to be embedded into the people's work and processes in order that it is fully exploited and the benefits reaped. Digital Enterprise Technology, a theoretical framework for collaborative design and production development in the context of a product's lifecycle, represents an emerging synthesis of new technologies. The framework can be used to configure digital product and process development technologies which assimilate design data, at various levels of completeness, with a high degree of real-time measurement feedback from the production environment in order to validate the product's tolerance specification and the selected production and assembly processes. It helps to reduce the implementation risk and streamline the assembly and integration processes.

Every activity/ process in the value chain uses some technology to combine raw materials or components and human resources to produce some output. Deployment of technology in the value chain would enable an organization to exploit the technological capabilities in operations. For example the adoption of enterprise resource planning (ERP) systems across the value chain (inbound logistics, operations and marketing and sales and distribution) can be used by a firm to not only improve its inventory and fixed assets turnover but also lead to efficiencies in marketing, sales and distribution. The customer relationship management (CRM) value chain helps to improve the efficiency of firms by organizing, aligning and integrating the organization processes along the value chain between the customer, the firm and its extended enterprise. (Chan, 2005). This is supplemented by customer intelligence wherein the firms use various technologies to collect data about customers from various sources spanning across all areas of an enterprise through many different business processes and involving many business units. (Chan, 2005).

Information Technology used in the various processes of the value chain affects positively the performance of the processes when it is not only strategically aligned to the business strategy of the firm but also that its primary locus (the point in the value chain where the impact of IT impacts are highest) is in processes that are considered essential for the execution of each business strategy. For example, for superior performance the locus of strategic alignment with information technology is in sales and marketing and customer relationships for customer intimate firms. (Tallon, 2007).

Therefore, selection of technology in this domain focuses not merely on the choice of specific technologies but also on the potential for technology integration. The extent to which the technology is adapted or altered so as to suit the needs of the organization also determines the level of operating efficiency attained by the organization. In fact, firms using very similar technologies might differ in the level of efficiency depending on the technology strategy followed after the initial adoption of technology.

Technical Support

Information technology used as a means of technical support can serve as an important means of cooperation between the different players across the value chain and the "end-users". One of such examples, the World Wide Web has emerged as an important tool to not only provide information to potential customers but also enable easy processing of electronic transactions. The use of the World Wide Web reduces search costs and the time required to search for potential suppliers.

"Transactional websites" include electronic catalogues, shopping carts, payment systems and order tracking systems that offer buyers a convenient and cost-effective way to support the procurement process by allowing for the identification and selection of suppliers and the execution of business transactions. (Benslimane, Plaisant & Bernard, 2005). Dramatic advances of information and communication technology as a means of technical support also has influenced the way organizations analyze customer information. Organizations are also utilizing IT interfaces to access histories of interactions with particular customers. This aggregate information has helped organizations to improve the quality of customer interactions.

Extensive knowledge (new as well as prior) is required at various stages of the production cycle. New software tools act as a means of capturing, representing and applying all types of knowledge for effective design and technical support required for production. According to Tatum (2005), timing of technical information is extremely critical for effective operations. Timely technical response to changed conditions and other field problems is an essential element for support.

INTERFACE BETWEEN INNOVATION AND TECHNOLOGY STRATEGY

The firm's innovative activities reflect the firm's technology strategy and the enactment of technology strategy serves to further develop its innovative capabilities. The technological innovation process can be broadly identified through four stages of problem recognition, technology selection, solution development and implementation while technology strategy can be understood through its substance and enactment. Successful technological innovation begins when a firm recognizes the potential of using technology to solve a particular problem or fulfill market needs.

This recognition has to a certain extent; roots in the business strategy of the firm which determines the extent to which technology can be used as a source of competitive advantage. During the technology selection stage the company evaluates different technologies which shall help the firm to come out with potential designs. The choice of a particular technology apart from being dependent on the extent the firm wants to use technology as a tool to compete also depends on the cost of technology, the resources the firm wants to invest in the technology and the diversity of technology portfolio which the firm feels competent to handle.

The solution development stage can be identified as the operationalization of the potential designs. This may proceed either by formulating a new solution from within the firm by investing in internal R&D activities or by adopting a ready-made solution from outside. This decision whether to develop solution internally or to source it from external sources, to a large extent is dependent on whether the firm wants to have a proprietary right on the solution developed which would give it a competitive edge as compared to other players in the market and also on how the firm evaluates the "make or buy" decision.

The full benefits of technological innovation are never fully realized until and unless the solution is implemented either in the form of new products in the market or in the form of cost reduction from improved processes. The strategy of an organization with respect to the deployment of technology drives the implementation of solution for technological innovation.

Taking this perspective, we can propose that technological innovation in an organization not only reflects the technology strategy of the organization but looking at technological innovation through the lens of substance and enactment of technology strategy we can propose the following. Apart from business strategy, the substance of technology strategy drives the technology selection stages of technological innovation, that is, in essence, the substance of technology strategy helps to identify what of technological innovation. The enactment of technology strategy by way of appropriation and deployment of technology reflects the solution development and implementation stages of technological innovation, wherein it helps to identify the how of technological innovation. The substance of technology strategy again drives the firm's decision with respect to development of a solution and implementation of the same. The implementation of technological innovation should serve as a source of feedback to the firm to refine and modify its technology strategy. Organizations do not lay down their technology strategies in isolation. The government and various policies and regulations are amongst the various environmental factors which affect the organizations' choice of technologies.

EFFECT OF GOVERNMENT, POLICIES AND REGULATIONS ON TECHNOLOGY STRATEGY

Government policy and various rules and regulations have an impact on the firm's technology strategy and its efforts towards innovation both in terms of funding and provision of guidelines for development. On a macro level, some of the critical factors are policies undertaken to promote macroeconomic stability; policies ensuring resource allocation in accordance with comparative advantage; rapid accumulation of physical and human capital; development of agricultural sector and promoting competent bureaucracies. (Westphal, 2002). In India there is a need to balance innovation policies that support traditional industry and technology with policies that better respond to issues of competition and enterprise development. (Chaturvedi, 2007). While the liberalization measures undertaken by the Government of India hit hard the Small and Medium Enterprises (SMEs) who found it hard to survive the competitive pressure, the 2006 budget of the Government introduced a special provision for improving credit flows to the SMEs.(Venkataramanaiah & Parashar, 2007). For example in China, the Government is committed to drive a number of regulatory reforms and institutional changes which are required to enhance Chinese knowledge creation and innovative capabilities. This gets manifested in the increase in R&D spending. Above all, the 'open policy' practiced by the Government remains the real key to the change in growth dynamics and innovation. (Simon, 2007).

The regulatory barriers to entry also play an important role in determining the dominance of firms in certain sectors. (Chataway, Tait & Wield, 2007). Strengthening of patent laws also plays an important role in transforming organizations from imitators to innovators. (Kale. & Little, 2007). The increasing importance of environmental issues leads to technological change and organizational change to enact it. (Boden, 1994).

IMPACT OF ORGANIZATIONAL FACTORS ON INNOVATION AND TECHNOLOGY STRATEGY

A review of literature on innovation has brought out the importance of various organizational factors which create a learning environment which promotes innovation.

Transition to a new technology within the organization faces not only financial barriers but a lot of cultural and political barriers too. Process technologies are intimately linked not only to the product technologies but also to organizational factors. (Rycraft, 2006). An organizational environment and culture which promotes learning and development of employees; open communication channels and learning from customers, suppliers and even competitors is in a better position to innovate. It is through learning that an organization is able to increase the depth and diversity of knowledge. In fact, higher the learning ability of a firm, higher is the level of company's competitiveness, innovativeness and product introduction success. (Yeung, 1999; Morales, Moreno.& Montes, 2007.).

Wilson (2007) introduced a new perspective of learning: learning from and learning with each other. In the former the stakeholders learn things that are already known to others from whom they are learning. In other words this involves a recycling of existing knowledge. Learning with is a collaborative and active process which involves creation of new knowledge. Such collaborations and networks not only promote continuous flow of information but also promote trust and reciprocal relationships which are valuable for rapidly innovating knowledge intensive technologies. (Rycroft, 2006).

Training and development programmes though not considered as innovative inputs, could be explicitly regarded as one of the main channels to upgrade the technological capabilities of firms. (Sirilli & Evangelista, 1998). The problem solving capabilities of knowledge workers lie in their education background, professional training, creativity and motivation. Approved and focused training programs help in production of new knowledge which thereby leads to innovative solutions and management of change. (Egbu, 2006). Moreover, due to change of technology from manual work to a high grade of automation due to advanced technology there has been an increase in demand for highly skilled work force. (Chroneer & Stenlund, 2006; Sohail. 1997).

Training of employees and development of champions also enables advanced tools to become embedded in the products and processes of the organization. (Barczak, Sultan & Hultink, 2007) Irrespective of organization's strategy to develop technology inside or to source from outside, it should provide users with sufficient and appropriate education and training on the new technology so as to make the adoption sustainable. Change in the mind-set of employees from "Not Invented Here Syndrome" to "Invented Anywhere Syndrome" also promotes harnessing of external technology for innovation.

(Witzeman, Slowinsky, Dirkx, Gollob, Tao, Ward & Miraglia, 2006). Technology deployed in the value chain requires inter-functional integration. Therefore organizational characteristics which encourage job rotation and inter-divisional teams will ensure that all the employees are seamlessly integrated in a supply chain and are constantly learning with an aim to promote innovation. These teams develop a tacit knowledge of how to handle critical situations. Shared between its members, this tacit knowledge may also determine the efficiency of the company. (Chroneer & Stenlund, 2006).

Execution of technology strategy so as to lead to innovation requires change in the set procedures in the organization. This upsets the status quo and often invites resistance from the employees. Cross functional communication and cooperation enables a change in the mindset of employees and makes them more flexible to changes. (Calabrese, 1999). It also promotes cross-disciplinary learning both within and across boundaries of the firm. Heterogeneity of knowledge, know-how and expertise available can increase the creativity level of employees. (Rodan & Galunic, 2004). Communication with stakeholders outside the organization also helps an organization to develop its dynamic capabilities which reflect an organization's capacity to develop new and innovative forms of competitive advantage and synergistic innovative capability. (Ayuso, Rodriguez & Ricart 2006; Persaud 2005). Organizational learning takes place only when knowledge is transferred throughout the organization, integrated with other knowledge areas and applied to a new product or process. (Kessler, Bierly & Gopalakrishnan, 2000). Organizational knowledge which is embedded in the interactions between employees of the firm and also between employees and external stakeholders are an important source for project ideas that provide input for technology strategy. (Meyer, 2008).

Additionally incentive schemes and rewards for sharing knowledge are important mechanisms to encourage employees to share information and experiences with each other. It not only enhances the organizations knowledge base but also enhances the spirit of teamwork. It also helps employees to understand where they fit into the collective dimension of the workplace. (Hsu 2006; Rezgui 2007). Incentive and performance measurement systems that reward cooperation and collective achievements instead of individual performance, only will lead to continuous enrichment and application of individual know-how. (Harryson,1997).

Another important organizational factor which has a profound impact on the innovative activity is the overall organization structure. In addition to the traditional hierarchical structures and the flat team based structures which promote formal relationships, there is an increasing trend towards informal relationships which span beyond organizational boundaries. (Wang & Ahmad, 2003). Organizations, should no longer guard their knowledge in pyramidal structures, but should form learning networks which span geographical locations and organizational boundaries. This would enable them to be more agile and responsive to innovations. "What organizations need is a willingness to experiment with alternative schemes instead of the rigid hierarchical structure of most organizations." (Goldman, 1985, p.8). According to (Fiol & Lyles, 1985) a centralized organization structure inhibits the internal learning process and makes diffusion of knowledge difficult across the organization. (as cited in Kessler, Bierly & Gopalakrishnan, 2000).

Formalization refers to the extent to which jobs within the organization are standardized. If a job is highly formalized the incumbent does not have much discretion regarding what job is to be done, how it is to be done or when it is to be done. Prior research Grover (1993) indicates that the degree of centralization is negatively related to innovation. High centralization of decision making leaves little autonomy in the hands of the team-members who thus resist any attempts of innovation by the organization.

Autonomy on the other hand, motivates the employees to try out new tools and techniques for the success of any product or process development. (as cited in Barczak, Sultan & Hultink, 2007). A consensus driven structure where horizontal and vertical communication is required deters implementation of risk-taking strategies whereas a hierarchical structure where decision-making and strategy formulation power is concentrated in a few hands promotes a risk-taking culture. It also promotes the adoption of radical process innovations. (Hemmert, 2008; Ettlie, Bridges & O'Keefe, 1984).

The decisions relating to technology strategy are usually the domain of the top management. (Sohail, 1997). It is the top management which plays a very important role in building a culture of learning and innovation. The attitudes, backgrounds and personalities of the Chief Executive and the group of senior executives who surround him determine the culture, traditions and personality of the corporation. "When the top management is rigid, conservative, driven by the numbers and the ROIs, one is not likely to see the element of risk taking that an innovative company demands." (Goldman, 2005, p.9). Innovation leadership requires a lot of communication and convincing until the vision has been absorbed throughout the organization. The presence of champions of particular technologies also determines the significant adoption of radical innovation processes. (Ettlie, Bridges & O'Keefe, 1984).

Effective leaders also play an important role in encouraging the organization to incorporate external sourcing of technology for innovation into the very fabric of the firm. They articulate visionary goals of complementing internal sources with external sources to achieve innovation and growth. These leaders reward employees for effective use of external sources. (Witzeman, Slowinsky, Dirkx, Gollob, Tao, Ward & Miraglia, 2006).

This implies "culture", a soft side needs to be developed along with formal strategies and structures within firms. Burgelman & Rosenblooms (1989), evolutionary process framework for technology strategy represents strategy making as a social learning process wherein technology strategy is inherently a function of the quantity and quality of organizational capabilities. (Hampsum & Tatum, 1997).

A CONCEPTUAL FRAMEWORK

Based on the above discussions, a model is being proposed. This model explains the relationship between technology strategy, technological innovation and organizational factors. Technological innovation in an organization is driven by its technology strategy which has its roots in the business strategy of the firm. Feedback obtained from implementation of technological innovation and the evolution of technology should form the basis for the organization to revise its technology strategy. The role of the Government as well as the various directives has an impact on the firm's technology strategy.

Organizational factors play a critical role in creating a learning environment to promote innovation. A bi-directional arrow between technology strategy and organizational factors, anticipate that it is not only the firm's technology strategy which has an effect on the various organizational factors but various factors like the expertise level and accumulated experience of employees play an important role with respect to the firm's choice of technology for implementation. The outcome effects of technological innovation can be measured as organizational performance with respect to its operations and business impact with respect to profit.

CONCLUSION

This chapter has proposed a model for exploring the relationship between technology strategy and technological innovation. Technology strategy cannot work in isolation to lead to innovation. It should be complemented by various organizational factors for competitive advantage.

Technology strategy of an organization can be understood by analyzing the technological innovation process. The paper highlights the importance of combining strategic and operational levels of analysis.

This model would help managers to manage their innovations better by pursuing appropriate technology strategy for innovation. HR and training & development policies can be modulated so as to support the technology strategy for innovations. A synergy between management of technology and management of softer aspects would benefit the organization.

However, since the model has been proposed on the basis of literature review it needs to be elaborated and refined through systematic research.

6

System Design in Traditional Concepts and Modern Paradigms

INTRODUCTION

System design is undergoing a series of radical transformations to meet performance, quality, safety, cost and time-to-market constraints introduced by the pervasive use of electronics in everyday objects. An essential component of the new system design paradigm is the orthogonalization of concerns, i.e., the separation of the various aspects of design to allow more effective exploration of alternative solutions. In particular, the pillar of the design methodology that we have proposed over the years is the separation between:

- Function (what the system is supposed to do) and architecture (how it does it);
- Communication and computation.

The mapping of function to architecture is an essential step from conception to implementation. When mapping the functionality of the system to an integrated circuit, the economics of chip design and manufacturing are essential to determine the quality and the cost of the system. Since the mask set and design cost for Deep Sub-Micron implementations is predicted to be overwhelming, it is important to find common architectures that can support a variety of applications [1].

To reduce design costs, re-use is a must. In particular, since system designers will use more and more frequently software to implement their products, there is a need for design methodologies that allow the substantial re-use of software. This implies that the basic architecture of the implementation is essentially "fixed", i.e., the principal components should remain the same within a certain degree of parameterization.

For embedded systems, which we believe are going to be the dominant share of the electronics market, the "basic" architecture consists of programmable cores, I/O subsystem and memories. A family of architectures that allow substantial re-use of software is what we call a hardware platform. We believe that hardware platforms will take the lion's share of the IC market.

However, the concept of hardware platform by itself is not enough to achieve the level of application software re-use we are looking for. To be useful, the hardware platform has to be abstracted at a level where the application software sees a high-level interface to the hardware that we call Application Program Interface or API. There is a software layer that is used to perform this abstraction. This layer wraps the different parts of the hardware platform: the programmable cores and the memory subsystem via a Real-Time Operating System (RTOS), the I/O subsystem via the Device Drivers, and the network connection via the network communication subsystem. This layer is called the software platform. The combination of the hardware and the software platforms is called the system platform.

In this paper, we first review the principles of system design, then offer a rigorous definition of platforms and show a methodology for their selection and use. We point to a forthcoming book for a detailed discussion of platform-based design using a less specific and formal view[1].

SYSTEM DESIGN PRINCIPLES

The overall goal of electronic system design can be summarized as follows:

Minimize

- Production cost,
- Development time and cost subject to constraints on performance and functionality of the system.

Production cost

Manufacturing cost depends mainly on the hardware components of the product. Minimizing production cost is the result of a balance between competing criteria. If we think of an integrated circuit implementation, then the size of the chip is an important factor in determining production cost. Minimizing the size of the chip implies tailoring the hardware architecture to the functionality of the product. However, the cost of a state-of-the-art fabrication facility continues to rise: it is estimated that a new 0.18?m high-volume manufacturing plant costs approximately $2-3B today. This increasing cost is prejudicing the manufacturers towards parts that have guaranteed high-volume production form a single mask set (or that are likely to have high volume production, if successful.) This translates to better response time and higher priorities at times when global manufacturing resources are in short supply.

In addition, the NRE costs associated with the design and tooling of complex chips are growing rapidly. The ITRS predicts that while manufacturing complex System-on-Chip designs will be practical, at least down to 50nm minimum feature sizes, the production of practical masks and exposure systems will likely be a major bottleneck for the development of such chips. That is, the cost of masks will grow even more rapidly for these fine geometries, adding even more to the up-front NRE for a new design. A single mask set and probe card cost for a state-of-the-art chip is over $1M for a complex part today, up from less than $100K a decade ago (note: this does not include the design cost). In addition, the cost of developing and implementing a comprehensive test for such complex designs will continue to represent an increasing fraction of a total design cost unless new approaches are developed.

As a consequence of this evolution of the Integrated Circuit world, if we determine a common "hardware" denominator (which we can call for now, platform) that could be shared across multiple applications, production volume increases and overall costs may eventually be (much) lower than in the case when the chip is customized for the application.

Of course, the choice of the platform has to be based on production volume but we cannot just forget the size of the implementation since a platform that can support the functionality and performance required for a "high-end" product may end up being too expensive for other lower complexity products. Today the choice of a platform is more an art than a science. We believe that a system design methodology must assist designers in this difficult choice with metrics and with early assessments of the capability of a given platform to meet design constraints (see for a comprehensive discussion of metrics to be adopted for platform-based design).

Development Cost & Time

As the complexity of the products under design increases, the development efforts increase exponentially. Hence to keep these efforts in check a design methodology that favors re-use and early error detection is essential. In addition, development time must be shorter and shorter to meet time-to-market requirements.

Short development times often imply changes in specifications while the product is being designed. Hence, flexibility, i.e., the capability of the platform to adapt to different functionalities without significant changes, is a very important criterion to measure the quality of a platform. Note that flexibility and re-use are related: the more flexible a platform is, the more re-usable. In addition to the ability to cover different behaviors, another important quality of a platform is the range of performance.

The flexibility/performance trade-off is very complex and it is strictly related to the ability of the platform to match a set of applications. For example, a hardware block implementing a digital filter with hardwired taps is not flexible since it cannot perform any other operation. The same function of digital filtering can be achieved using a software-programmable Digital Signal Processor (DSP) that is clearly much more flexible since it can perform a large set of operations. Limiting, for sake of simplicity, the set of behavior to the class of finite impulse response (FIR) digital filters, the length of the filter response or number of taps characterize the filtering function, the maximum sampling frequency its performance. The filtering function implemented as a hardware block is limited to a fixed number of taps, while the filtering function implemented on a DSP can have a variable number of taps, only limited by the available memory of the DSP. On the other hand, the performance range for the hardware implementation of the digital filter is fairly large, while for the DSP implementation is lower and decreases with the number of taps.

Flexibility

The most flexible platforms include software programmable components. In this case, flexibility is achieved using a "machine" able to perform a set of instructions specified by the instruction set architecture (ISA). The execution of the sequence of instructions, or instruction stream stored in memory, realizes the desired behavior (Processor unit, micro-code unit). Powerful 32-bit microprocessors are very flexible since they can perform an almost unlimited number of functions with good performance, small 8-bit microprocessors have the same functional flexibility but their performance is clearly much more limited.

An intermediate point between software programmable components and hardware blocks in the space of flexible objects consists of re-configurable parts. Re-configurable components are characterized by hardware that can change the logic function it implements by changing the interconnection pattern among basic building blocks and the function implemented by the basic building blocks themselves. For some re-configurable components this change can be done only once at design time (for example, FPGAs with anti-fuses and EPROMs), for some, it can be done many times even at run-time (FPGAs with RAM-like cells, EEPROMs, Flash EEPROMs). There are re-configurable components where the basic building blocks have large granularity and re-configuration is achieved by changing appropriate parameters in memory. An example is the run-time re-configurability of microprocessor-based systems where peripherals can be connected with different patterns according to the application. Re-configurable components have been used for years not only for fast-prototyping but also for final products in the initial stages of their market introduction. Because of their superior performance in terms of speed and power consumption with respect to instruction-set machines, re-configurable components are a valid choice for several systems. Research has been carried out to identify novel re-configurable architectures that are more flexible and more performing than the present choices [3].

Design Re-use

Design re-use can be achieved by sharing components among different products (re-use in space) and across different product generations (re-use in time). Re-use-in-time has to bridge the technology gap between two (or even more) generations. Re-use-in-space is based on commonality among different products.

It depends on the capability of designers to partition their designs so that similarities among different products are factored as common terms. Both re-use and early error detection imply that the design activity must be defined rigorously so that all phases are clearly identified and appropriate checks are enforced. To be effective, a design methodology that addresses complex systems has to start at high levels of abstraction.

In several system and IC companies, designers are used to working at levels of abstraction that are too close to implementation so that sharing design components and verifying before prototypes are built is nearly impossible. Design methodologies that address this problem emerged recently and design tools and environments can be put in place to help supporting this design methodology (see the Polis and VCC Felix design systems and the methodology they support) [2][6].

Design re-use is most effective in reducing NRE costs and development time when the components to be shared are close to final implementation. For hardware components, re-use at the implementation level means re-use of physical components avoiding new mask design and production. On the other hand, it is not always possible or desirable to share designs at this level since minimal variations in specification may result in different, albeit similar, implementations.

The ultimate goal is to create a library of functions and of hardware and software implementations that can be used for all new designs of a company. It is important to have a multi-level library since it is often the case that the lower levels that are closer to the physical implementation change due to the advances in technology while the higher levels tend to be stable across product versions.

Design re-use is desirable both at hardware and software level:

- By fixing the hardware architecture, the customization effort is entirely at the software level. This solution has obvious advantages in terms of design cycles since hardware development and production cycles are indeed longer than their software counterpart.
- Basic software such as RTOS and device drivers can be easily shared across multiple applications if they are written following the appropriate methodology. This is true in particular for device drivers as will be discussed in details later.
- Application software can be re-used if a set of conditions are satisfied:
- Each function is decomposed in parts with the goal of identifying components that are common across different products. In the best

case, the entire functionality can be shared. More often, sub-components will be shared. In this case, the decomposition process is essential to maximize design re-use. The trade-off here is the granularity of the components versus the sharing potential. The smaller the parts are, the easier is to share them but the smaller is the gain. In addition, sharing does not come for free. Encapsulation is necessary for both hardware and software to prevent undesired side effects. Encapsulation techniques have been extensively used in large software designs (object orientation) but only recently have caught the attention of embedded system designers [7].

- "High-level" languages (e.g., C, C++, Java) are easily re-targetable to different processors. However, standard compilers and interpreters for these languages are not efficient enough for the tight requirements that system designers must satisfy. Hence they often exploit the micro-architecture of the processor to save execution time and memory, thus making the re-usability of their software across different microprocessors almost impossible. If super-optimized object code generation with the execution time versus memory occupation trade-off made visible to the designers were available, then re-usability would be finally easy to achieve. Our research groups and others have attacked this very topic to allow embedded system designers to focus on the high-level aspects of their problem (see the code generation techniques of Polis[6], software synthesis of data flow graph [8] and super-compilation for VLIW architecture [9]).

DESIGN METHODOLOGY

Once the context of the design methodology is set, we can move on to define some important concepts precisely that form the foundation of our approach.

Function

A system implements a set of functions. A function is an abstract view of the behavior of the system. It is the input/output characterization of the system with respect to its environment. It has no notion of implementation associated to it. For example, "when the engine of a car starts (input), the display of the number of revolutions per minute of the engine (output)" is a function, while "when the engine starts, the display in digital form of the number of revolutions per minute on the LCD panel" is not a function. In this case, we already decided that the display device is an LCD and that the format of the data is digital. Similarly, "when the driver moves the direction indicator (input), the display of a sign that the direction indicator is used until it is returned in its base position" is a function, while "when the driver moves the direction indicator, the emission of an intermittent sound until it is returned to its base position" is not a function.

The notion of function depends very much on the level of abstraction at which the design is entered. For example, the decision whether to use sound or some other visual indication about the direction indicator may not be a free parameter of the design. Consequently, the second description of the example is indeed a function since the specification is in terms of sound. However, even in this case, it is important to realize that there is a higher level of abstraction where the decision about the type of signal is made. This may uncover new designs that were not even considered because of the entry level of the design. Our point is that no design decision should ever be made implicitly and that capturing the design at higher levels of abstraction yields better designs in the end.

The functions to be included in a product may be left to the decision of the designer or may be imposed by the customer. If there are design decisions involved, then the decisions are grouped in a design phase called function (or sometimes feature) design. The decisions may be limited or range quite widely.

Architecture

An architecture is a set of components, either abstract or with a physical dimension, that is used to implement a function. For example, an LCD, a physical component of an architecture, can be used to display the number of revolutions per minute of an automotive engine. In this case, the component has a concrete, physical representation. In other cases it may have a more abstract representation. In general, a component is an element with specified interfaces and explicit context dependency. The architecture determines the final hardware implementation and hence it is strictly related to the concept of platform.

The most important architecture for the majority of embedded designs consists of microprocessors, peripherals, dedicated logic blocks and memories. For some products, the architecture is completely or in part fixed. In the case of automotive body electronics, the actual placement of the electronic components inside the body of the car and their interconnections is kept mostly fixed, while the single components, i.e., the processors, may vary to a certain extent. A fixed architecture simplifies the design problem a great deal but limits design optimality. The trade-off is not easy to achieve.

We call an architecture platform, a fixed set of components with some degrees of variability in the performance or other parameters of one or more of its components.

Mapping

The essential design step that allows moving down the levels of the design flow is the mapping process, where the functions to be implemented are assigned (mapped) to the components of the architecture. For example, the computations needed to display a set of signals may all be mapped to the same processor or to two different components of the architecture (e.g., a microprocessor and a DSP). The mapping process determines the performance and the cost of the design.

To measure exactly the performance of the design and its cost in terms of used resources, it is often necessary to complete the design, leading to a number of time consuming design cycles.

This is a motivation for using a more rigorous design methodology. When the mapping step is carried out, our choice is dictated by estimates of the performance of the implementation of that function (or part of it) onto the architecture component. Estimates can be provided either by the manufacturers of the components (e.g., IC manufacturers) or by system designers.

Designers use their experience and some analysis to develop estimation models that can be easily evaluated to allow for fast design exploration and yet are accurate enough to choose a good architecture. Given the importance of this step in any application domain, automated tools and environments should support effectively the mapping of functions to architectures [2][6].

The mapping process is best carried out interactively in the design environment. The output of the process is either:

- A mapped architecture iteratively refined towards the final implementation with a set of constraints on each mapped component (derived from the top-level design constraints) or
- A set of diagnostics to the architecture and function selection phase in case the estimation process signals that design constraints may not be met with the present architecture and function set. In this case, if possible, an alternative architecture is selected. Otherwise, we have to work in the function space by either reducing the number of functions to be supported or their demands in terms of performance.

Link to Implementation

We enter this phase once the mapped architecture has been estimated as capable of meeting the design constraints. We now have the problem of implementing the components of the architecture. This requires the development of an appropriate hardware block or of the software needed to make the programmable components perform the appropriate computations. This step brings the design to the final implementation stage. The hardware block may be found in an existing library or may need a special purpose implementation as dedicated logic.

In this case, it may be further decomposed into sub-blocks until either we find what we need in a library or we decide to implement it by "custom" design. The software components may exist already in an appropriate library or may need further decomposition into a set of sub-components, thus exposing what we call the fractal nature of design, i.e., the design problem repeats itself at every level of the design hierarchy into a sequence of nested function-architecture-mapping processes.

HARDWARE PLATFORMS

Product features can be used as a metric for measuring the complexity of the functionality of a product. Figure shows six different products in the same product family ranked according to product features and lined up in time of introduction in the market place. We can group these products into two classes: high-end (A,B,C) and low-end (D,E,F), according to the number of features and their overall complexity.

If we use the same architecture for both classes, then the hardware part of the design is fully re-used while software may need to be partially customized.

If indeed the high-end class strictly contains the features (functions) of the other, then software design can also be in large part re-used. Thus even if unit cost were higher than needed for the low-class products, design costs and NREs may push towards this solution. We have witnessed this move towards reducing product variants in a number of occasions both in system and IC companies. The common architecture can then be identified as a platform. We feel that, given its great importance, a definition of hardware platform is needed since different interpretations have been used of this commonly used term.

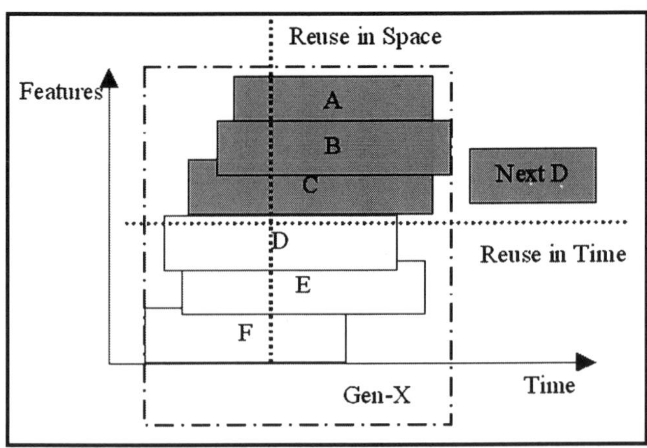

Fig. Product Space

A hardware platform is a rather warm and fuzzy concept that is related to a common architecture of sort. We believe that it is possible to generalize the basic idea to encompass not only a fully specified architecture but it is also a family of architectures that share some common feature. Hence we prefer to identify a platform with the set of constraints that can be used to test whether a given architecture belongs to the family.

The "strength" of hardware and software architectural constraints defines the level and the degree of re-use. The stronger the architectural constraints the more component re-use can be obtained. On the other hand, stronger constraints imply also fewer architectures to choose from and, consequently, less application-dependent potential optimization.

The Personal Computer (PC) [5] platform is a good example of tight hardware and software constraints that enable design re-use (both in space and in time). The essential constraints that determine the PC hardware platforms are:

- The x86 instruction set architecture (ISA) that makes it possible to re-use the operating system and the software application at the binary level;
- A fully specified set of busses (ISA, USB, PCI) that make it possible to use the same expansion boards or IC's for different products;
- Legacy support for the ISA interrupt controller that handles the basic interaction between software and hardware.
- A full specification of a set of I/O devices, such as keyboard, mouse, audio and video devices.

All PCs satisfy this set of constraints. The degrees of freedom for a PC maker are quite limited! However, without these constraints the PC will not be where it is today. One wonders why this success has not percolated in the embedded system domain! This question will be answered in the following paragraphs.

An embedded system hardware platform is a combination of three main sub-systems: processing unit, memory, and I/O.

- Strong cost and packaging requirements impose keeping the hardware components at minimum complexity and size, thus making the adoption of the platform concept more difficult.
- The interaction with the environment is frequently very complex and application dependent. Hence, the I/O subsystem is an integral and essential part of the architecture as it cannot be assumed to be equal across the application space as is the case for the PC domain. Consequently, the architecture itself is application specific and not re-usable across different application domains. Even for the same application domain, embedded systems can differ very much in their I/O sub-systems. In the car dashboard example, low-end and high-end products have completely different I/O requirements: the latter use very complex display and communication devices while the former use simple display and serial communication.

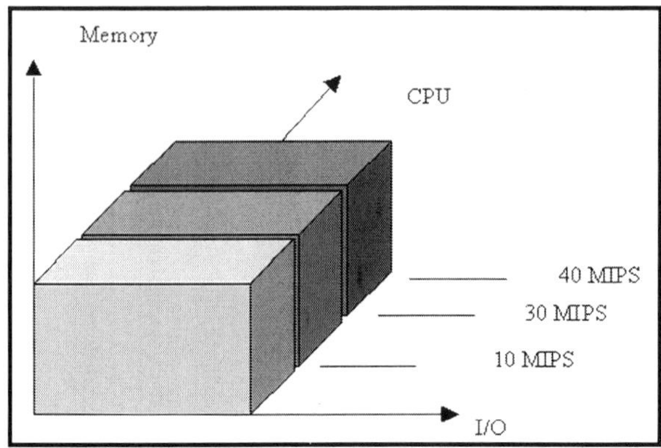

Fig. Example of platform space

Figure above shows different versions of an embedded system platform parameterized with respect to:

- Processing unit performance;
- Speed and footprint for the memory system (RAM, ROM, Flash, EEPROM...);
- Analog channels, timing and digital channels, communication speed for the I/O sub-system.

Note that the hardware platform is a family of possible products fully identified by the set of constraints. When we define a product, we need to fully specify its components. The six products requirements covered by two different instances of a unique hardware platform: high and low end. The I/O dimension is a projection of a very complex space where characteristics such as number of digital I/Os, number of PWM channels, number of A/D converters, number of input captures and output compares, and specific I/O hardware drivers are represented.

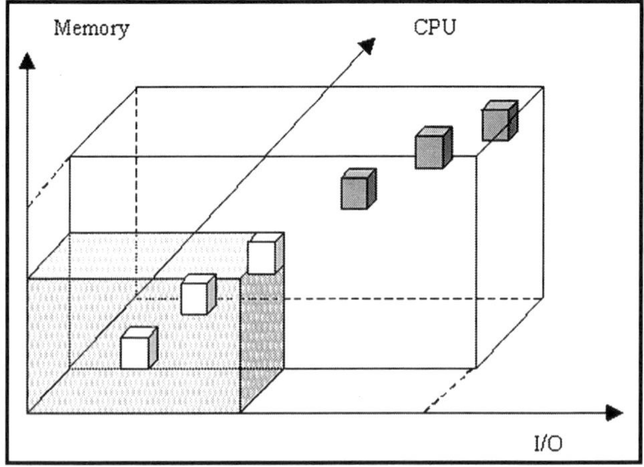

Fig. Platforms for different product classes

The two hardware platform instances are also shown in the features/time space. The diversification of hardware architectures should be carefully handled since it reduces the production volume of each diversification.

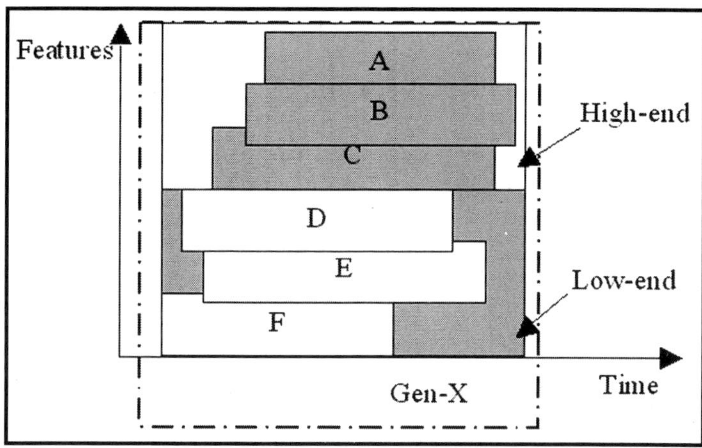

Fig. Platforms in the feature/time space

Seen from the application domain, the constraints that determine the hardware platform are often given in terms of performance and "size". For the dashboard example, we require that, to sustain a set of functions, a CPU should be able to run at least at a given speed and the memory system should be of at least a given number of bytes. Since each product is characterized by a different set of functions, the constraints identify different hardware platforms where more complex applications yield stronger architectural constraints. Coming from the hardware space, production and design costs imply adding hardware platform constraints and consequently reducing the number of choices.

The intersection of the two sets of constraints defines the hardware platforms that can be used for the final product. Note that, as a result of this process, we may have a hardware platform instance that is over-designed for a given product, that is, some of the power of the architecture is not used to implement the functionality of that product. Over-design is very common for the PC platform. In several applications, the over-designed architecture has been a perfect vehicle to deliver new software products and extend the application space. We believe that some degree of over-design will be soon accepted in the embedded system community to improve design costs and time-to-market. Hence, the "design" of a hardware platform is the result of a trade-off in a complex space that includes:

- The size of the application space that can be supported by the architectures belonging to the hardware platform. This represents the flexibility of the hardware platform;
- The size of the architecture space that satisfies the constraints

embodied in the hardware platform definition. This represents the degrees of freedom that architecture providers have in designing their hardware platform instances.

Once a hardware platform has been selected, then the design process consists of exploring the remaining design space with the constraints set by the hardware platform. These constraints can not only be on the components themselves but also on their communication mechanism. When we march towards implementation by selecting components that satisfy the architectural constraints defining a platform, we perform a successive refinement process where details are added in a disciplined way to produce a hardware platform instance.

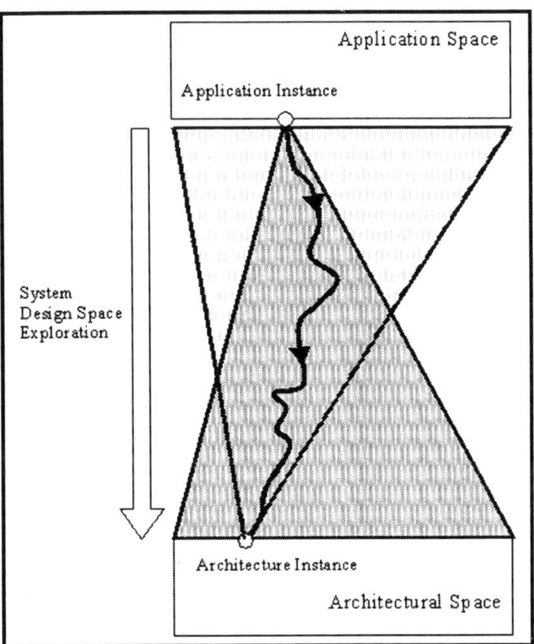

Fig. Architectural space implementing the application instance (at different cost).

Ideally the design process in this framework starts with the determination of the set of constraints that defines the hardware platform for a given application. In the case of a dashboard, we advocate to start the design process before splitting the market into high-end and low-end products. The platform thus identified can then be refined towards implementation by adding the missing information about components and communication schemes. If indeed we keep the platform unique at all levels, we may find that the cost for the low-end market is too high. At this point then we may decide to introduce two platform instances differentiated in terms of peripherals, memory size and CPU power for the two market segments. On the other hand, by defining the necessary constraints in view of our approach, we may find that a platform exists that covers both the low-end and the high-end market with great design cost and time-to-market improvements.

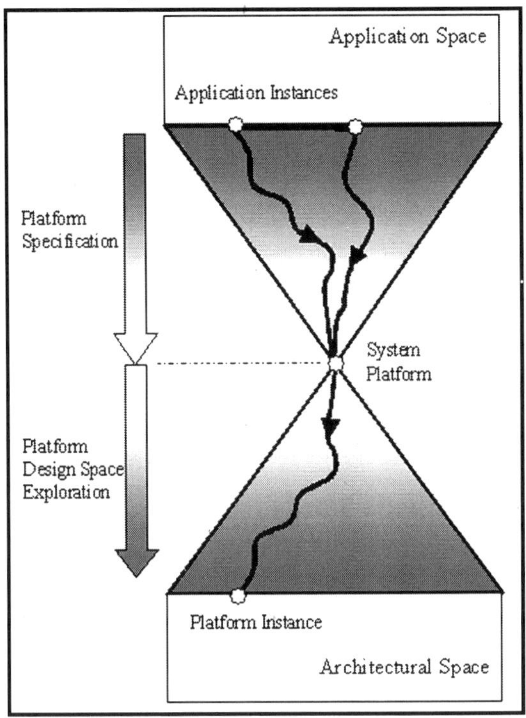

Fig. The system platform concept forces the design exploration to find somewhere in the abstraction a common solution.

Hardware platform-based design optimizes globally the various design parameters including, as a measure of optimality, NRE costs in both production and design. Hardware platform-based design is neither a top-down nor a bottom-up design methodology. Rather, it is a "meet-in-the-middle" approach. In a pure top-down design process, application specification is the starting point for the design process. The sequence of design decisions drives the designer toward a solution that minimizes the cost of the architecture. The single application approach, the bottom of the figure shows the set of architectures that could implement that application. The design process, the black path, selects the most attractive solution as defined by a cost function. In a bottom-up approach, a given architecture (instance of the architectural space) is designed to support a set of different applications that are often vaguely defined and is in general much based on designer intuition and marketing inputs.

- An essential feature to achieve re-usability of a hardware platform is that the application designer sees it as a unique object in his design space. For example, in the PC world, the application software designer sees no difference among the different PCs that are satisfying the PC platform architectural constraints.

Figure shows graphically the features of an hardware platform. The cone above the point represent its flexibility, the cone below represents its generality. In general, hardware platforms tend to have a large cone above and a small cone below to witness how important design re-use and standards are versus the potential optimization offered by loose constraints.

SOFTWARE PLATFORM

The concept of hardware platform by itself is not enough to achieve the level of application software re-use we are looking for. To be useful, the hardware platform has to be abstracted at a level where the application software "sees" a high-level interface to the hardware that we call Application Program Interface or API. There is a software layer that is used to perform this abstraction. This layer wraps the essential parts of the hardware platform:

- The programmable cores and the memory subsystem via a Real-Time Operating System (RTOS),
- The I/O subsystem via the Device Drivers, and
- The network connection via the network communication subsystem.

This layer is called the software platform.

In our conceptual framework, the programming language is the abstraction of the ISA, while the API is the abstraction of a multiplicity of computational resources (concurrency model provided by the RTOS) and available peripherals (Device Drivers). There are different efforts that try to standardize the API [12], [13], [14]. In our framework, the API is a unique abstract representation of the hardware platform. With an API so defined, the application software can be re-used for every platform instance.

Of course, the higher the abstraction layer at which a platform is defined, the more instances it contains. For example, to share source code, we need to have the same operating system but not necessarily the same instruction set, while to share binary code, we need to add the architectural constraints that force to use the same ISA, thus greatly restricting the range of architectural choices.

RTOS

In our framework, the RTOS is responsible for the scheduling of the available computing resources and of the communication between them and the memory subsystem. Note that in most of the embedded system applications, the available computing resources consist of a single microprocessor. However, in general, we can imagine a multiple core hardware platform where the RTOS schedules software processes across different computing engines. There is a battle taking place in this domain to establish a standard RTOS for embedded applications. For example, traditional embedded software vendors such as ISI and WindRiver are now competing with Microsoft that is trying to enter this domain by offering Windows CE, a stripped down version of the API of its Windows operating system. In our opinion, if the conceptual framework we offer here is accepted, the precise definition of the hardware platform and of the API should allow to synthesize automatically and in an optimal way most of the software layer, a radical departure from the standard models borrowed from the PC world. Software re-use, i.e. platform re-targetability, can be extended to these layers (middle-ware) hopefully resulting more effective than binary compatibility.

Device Drivers

The "virtualization" of input and output devices has been the golden rule to define generic and object-oriented interfaces to the real word. The I/O subsystem is the most important differentiator between PCs and embedded systems.

In our framework, the communication between software processes and the environment, i.e. physical processes, is refined (input/output refinement) through different software and hardware layers. The device drivers identify the software layers. The layer structure is introduced to achieve reusability of components belonging to each layer.

In this scheme, the acquisition of a physical variable is seen as a communication between a sender, that "lives" in the physical domain, and a receiver, that "lives" in the virtual domain. At the application level, the acquired value is an image, i.e. a representation with a certain precision and time validity, of the physical variable correctly sampled and quantized. No details about the sensor, analog-to-digital converter and other hardware components are known.

Using the analogy with protocols, to implement this communication three main layers are identified: the presentation, transport and physical layer. For the sensor, i.e. the sender, the presentation and transport layers serve as a "transducer" of the physical variable to an electrical variable, which is transported, with wires, to the hardware of the system. I/O communication is asymmetric, i.e. the communication behavior is typically sampling and quantization for the link from the environment to the system and hold for the opposite direction.

The hardware channels are abstracted via a software layer that "lives" underneath the device driver: the Basic I/O system or BIOS. This software layer must guarantee the reliability of the electrical link by continuously monitoring the link and its quality. The BIOS represents the transport layer that typically can support different types of sensors/actuators satisfying bandwidth and accuracy of the I/O channel. Thus the layering of the device driver offers another opportunity of design re-use.

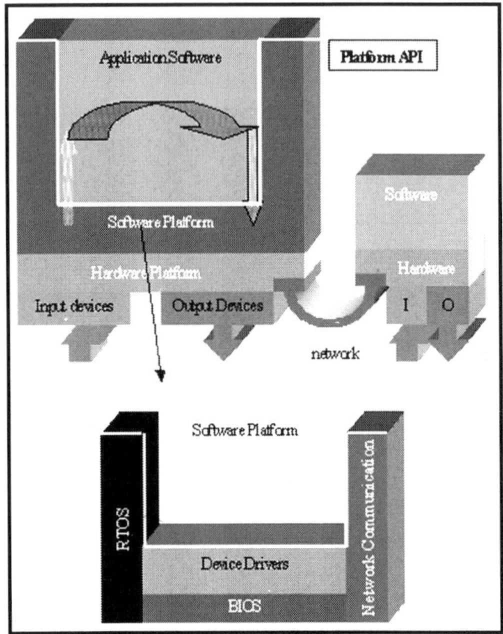

Fig. Layered software structure

For the receiver, i.e. the application process, the presentation layer masks the operations required to extract the image of the physical variable out of the electrical information provided by the transport layer. Typical operations are filtering, diagnosis and recovery of the sensor signal and sensor calibration.

In this communication abstraction, transducers implement the translation of the physical variable towards its electrical representation and device drivers represent its abstraction towards the application software with no attention towards the transport mechanism that is left to the BIOS.

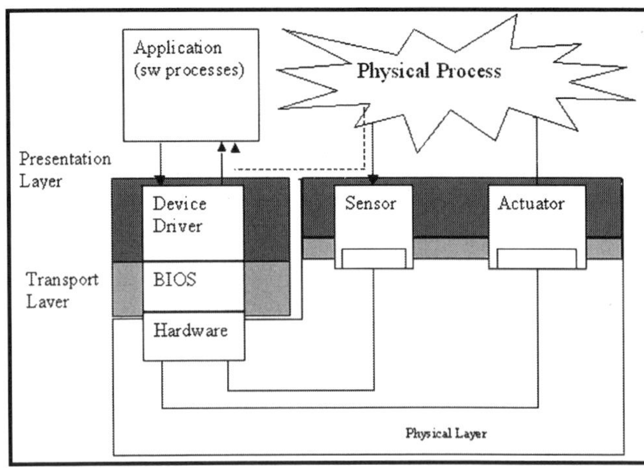

Fig. IO subsystem as a communication link

Network Communication

The network communication subsystem supports the connection between software processes running on different electronic systems. Information about the status of remote parts of the system is typically exchanged. In a car, in-vehicle communication is required to continuously update the status of physical variables acquired by a different electronic control unit, such as car speed, battery voltage or brake pedal status. In these cases, the network subsystem is the mean to "acquiring" a physical variable from remote systems but no difference is seen from the application software making re-use really effective.

SYSTEM PLATFORM

The combination of hardware and software platforms is called system platform. It is the system platform that completely determines the degree of re-usability of the application software and identifies the hardware platform. In fact, the hardware platform may be deeply affected by the software platform as in the case of the PC platform. In fact, the re-usability of the application software made possible by the PC system platform comes at the expense of a very complex layered structure of the Windows O.S. families and of the size and complexity of the microprocessor cores supporting the instruction set. In this domain, this price is well worth being paid.

In the automotive domain, the use of the OSEK/VDX RTOS [4] specifications, sponsored by the German automotive manufacturers, was conceived as a tool for re-usability and "flexibility" using the PC platform as a model. However, the peculiarities of embedded systems where cost, reliability and geometric constraints play a fundamental role, make this approach much more difficult to use. The OSEK initiative intends to set a standard for operating systems to be used in the car so that software produced by different vendors could be easily integrated.

OSEK constrains the RTOS and part of the network communication system, but does not constrain any input/output system interface exposing the hardware details of the I/O subsystem to the software. Because in the OSEK specification a device driver API is not defined, this specification is not an architectural constraint strong enough to enable alone the re-use of software components even at source code level, letting each system maker define its own interface to interact with the attached devices. If we add to the OSEK standard for RTOS and Network management, the definition of the I/O sub-systems and the use of a single ISA, even binary software components can be re-used.

On the other hand, OSEK fully specifies the scheduling algorithm for the operating system. It does guarantee that no priority inversion occurs but it does not help in a substantial way to make the verification of the timing properties of the application any easier. In our opinion, this is an over-specification with respect to software re-usability since, even if the scheduling algorithm is not fully specified, we can still re-use software at the source level. The jury is still out to determine whether the overall OSEK constraints are so strong that economic and efficient solutions are eliminated a priori. We doubt that this specification in its present form is going to yield the desired results just because of the difficult trade-off space that embedded applications entail.

A DEVELOPMENT PARADIGM?

After considering what has been discussed so far, the unresolved question that remains in our case would then be: Is there a dominant or monopolic paradigm in development work as we see it in the Western context? And further, is such a paradigm multidisciplinary or does each contributing discipline have its own paradigm? A dominant development paradigm would imply a shared definition of the field of knowledge that more or less uniformly affects the group that practices in that field. Under such a premise, it must be pointed out that most of us development workers are yet to be "convened" into such an emerging paradigm.

A missing paradigm in our analysis?:

Many of us in development work already are a mixture of natural and social scientists and, therefore, move inside at least two paradigms. Moreover. those among us who are strongly committed to a bottom-up, community-oriented approach in this work wonder if there is such a thing as a "people's paradigm" or a "community-to-be-developed paradigm." (If one looks at some of the definitions of paradigm here presented, i.e. those by Jevons, this should not be an outrageous proposition.) If the answer is yes, would that mean that we scientists need to always search for, define, and integrate this seemingly indispensable people's paradigm? Would the fact of not having done so for decades be where we have failed most precipitously, explaining why we have so often failed in our efforts to promote and bring about development? Is considering the integration of this people's paradigm the basis for a truly democratic approach to development...?

The discrepancy between our level of analysis and that of the beneficiaries of development interventions and the discrepancy between those analytical levels and our level of concrete actions or propositions for action in development matters is part of our paradigms' limitations that lock us in, consciously or not, most probably because of deep-rooted ideological barriers.

A case in point in the area of food and nutrition was the 1974 World Food Conference in Rome. It did not really politicize the issues as was probably needed, but it unified some paradigms. Now, everybody accepts - at an analytical level - that hunger is a social science concept and of social science's concern, especially when one looks at its economic.. and socio-political determinants and consequences (21). At the level of concrete actions to tackle the causes identified, however, little can be shown as concrete achievements 13 years after the Conference, thus illustrating the discrepancy just pointed out.

In the same above context, one wonders why have development theories only very secondarily touched on hunger and malnutrition issues? In more recent years, important development documents such as the Brandt Commission Report, the Scheweningen Report and the Lagos Plan of Action explicitly mention hunger and malnutrition, but resolving the latter has not been clearly made into an integral part of the solution, because the basic social injustices at its base have simply not been tackled decisively (22, 23). Another thing to bear in mind following this idea of the solutions falling short of verbal commitments is that, to begin with the modernization paradigm in development work has long assumed that societies characterized by industrial capitalism are universally desired (or desirable). In fact, no people have ever directly voted for capital accumulation and industrialization (24).

Paradigms and the ruling elites:

The trouble with us Western scientists and intellectuals is that while we engage in this never-ending discussion, paradigmatic pronouncements, and principles generated by us in the sciences are being taken advantage of on an everyday basis, by the pro-status quo ruling elites who have no real commitments to meaningful change to improve the quality of life of the lot. They, therefore, allocate no or highly insufficient budgets to real development priorities and to nobody's big surprise nothing much improves. In our lobbying with decision-makers we too often settle for only a threshold consensus, but incremental changes somehow always are taken advantage of by the elites, most often the local bourgeoisie.

Attempts at gradual repairs of the system are usually normative, often carried out in a token fashion and are sometimes plainly naive. Unfortunately reality seems to show that ideas born in diplomacy, stressing existing harmony rather than existing conflict (overt or underlying), are flawed and end up leaving things as they are (25).

Let's face it: The ruling class in capitalist-dependent countries is not ignorant, but rather reactionary. Therefore, little is accidental about the problems of under- and maldevelopment. The search for a consensus with decision makers ultimately impinges in what is ideologically licit or illicit. Therefore, improving the dialogue with them is not possible if the price is ducking the contradictions of the system; deep ideological (class) barriers make a breakthrough unlikely.

This inevitability means we need to prepare for some kind of conflict and confrontation. Solving long-standing contradictions and injustices takes a protracted time of negotiations and involves an organized struggle. In the last instance, the biggest aggressor and violator of human rights inherently is the capitalist system and that denunciation alone has no big publicity interest anymore and only few are willing to listen to it (25, 26).

MULTIDISCIPLINARITY

According to Webster's Dictionary (1), a discipline relates to a particular field of study or subject: it connotes a pattern of behavior and rules or a system of rules governing conduct. A multidisciplinary approach, by extension, would bring together persons with different backgrounds and from different disciplines.

Multidisciplinarity: A panacea in development work?

At the base of multidisciplinarity is the concept that complex problems have many facets that one or two persons with conventional training in one or two fields cannot handle appropriately. The ides is that the different approaches can be bridged by working on the problems together. Multidisciplinarity comes to us "as a natural" in development work when we see that communities have problems of many sorts and we in academia have encircled ourselves in rigid departments. The problem is that the understanding of what a multidisciplinary approach and its potential is and who the members of such a team should be is different for people with different ideological perspectives; it very much depends on how one slices reality into categories and disciplines (e.g., should political scientists be part of multidisciplinary development teams?). Scholars looking at the same problems from different angles often seem to be unable to listen and learn from each other.

A good example that comes to mind is the problem of hunger and malnutrition. It is still basically defined from the disciplinary bias of the scientists: for the agronomist it is mainly a food supply problem; for the educator, a problem of ignorance; for the demographer, a problem of population pressures; for the planner, it is the lack of coordination, and so forth. Most of these approaches are not wrong, but utterly incomplete or partial and have eventually led to generalizations that have served to create and spread myths that may in the end have the ring of truth about them, simply because they are repeated often enough (26, 27).

When involved in multidisciplinary work, we are forced to make both individual and collective decisions. We have to decide, for instance, which principles - or "unproblematic background knowledge" - we will take by fiat. The surrounding scientific community will always operate as a base for such a validation. The ultimate explanatory principles will thus be heavily ideological in their structure. Therefore, none of our decisions as team member can be made purely on the base of scientific arguments alone since it is our own scientific community that assures that "objectivity" (28, 29). Multidisciplinarity, as seen in Western development work, thus fosters an equilibrium-centered view that often emphasizes constancy in behavior over time. Praxis, more than scientific experiment, has exposed the fundamental paradox of fostering such a status quo through traditional multidisciplinary approaches.

Success in achieving constancy inevitably leads to crisis, more difficult and costly later problems and more unexpected future events. If multidisciplinary development work just espouses an evolutionary view, highlighting organizational changes rather than structural changes, understanding will be sought by analyzing parts of the whole rather than the whole. But if we are to understand the part we must understand the whole! Generality is difficult to capture in the face of multidisciplinarity's great diversity but it can be achieved by separating the causes of a process that are universal from those that are specific only to a subset of situations and then analyzing the combined effects of the latter (30).

A detailed explanation is not the same as an understanding and explanations is what multidisciplinary teams are mostly offering. Hence, for a long period, understanding has been sought largely by applying partial facets of knowledge in order to comprehend higher order phenomena. It is a view dominated by immediate cause/effect determinism, by a quantitative emphasis and by a paradigm of stability and order. As a case in point, malaria eradication programs have been brilliant examples of sophisticated understanding combined with a style of implementation that has all the character of a military campaign.

Where malaria was neither marginal nor at low endemic levels, however, transient success led to human populations with little immunity and mosquito vectors resistant to DDT endangering not only the health of the population, but also overall socioeconomic development. We can be successful in achieving short-term objectives, but as a consequence of this success, problems sooner or later evolve into qualitatively different ones, pointing to the more structural causes of the original problem. For too many of the last 20 years in development work emphasis has been on developing a qualitative understanding with a departure from quantitative explanations. Non-judicious application of quantitative techniques leads to an oversimplification of reality and to a historical analyses (30, 31).

Multidisciplinarity also has a tendency to fall into many of the intellectual traps that characterize complex systems, e.g., treating symptoms in isolation or seeking short-term political solutions. This process often involves a set of ameliorative policies, each of which is "rational" for the particular problem it is designed to solve, but which when aggregated either exacerbate the problems they were designed to solve or create a new set of problems. As development workers, we have the tendency to devise solutions based on a static analysis of a problem whereas it is the dynamic, time-dependent analysis that is by far the most important. As our alienation increases, a downward spiral in bad decisions leading to accumulation of small failures ensues; as system performance declines, the public will look for (and politicians, perhaps even through our advice will be tempted to offer) simple solutions based purely on gut responses. After all, ideology justifies any particular policy decision made. A conservative ideology-or a liberal one for that matter - often takes the complexity out of the problem. Fragmented policy responses, however, often result in action that is largely symbolic and not very effective in the long run (32).

Insofar as multidisciplinary teams in development work attend to the public dimension of their ethical responsibilities at all, at present, most tend to see their public duties as obligations to promote the public interest - associated with a vision of society as a rational alliance of primarily self interested persons whose own "good" is made up of a complex of private interests. The aggregation of the latter interests should lead to mutual advantage.

To promote the public interest is to maximize the collective realization of individual interests and "to protect the integrity and functioning of those social arrangements, institutions, and values that make orderly social life possible and mutually (...?) advantageous". But this is simply not enough! Important as they are, activities such as contributing to the analysis and to the making of public policy, as well as providing services to beneficiaries do not, in the aggregate, exhaust the duties that the multidisciplinary development team members ought to have. The public duty of our multidisciplinary teams extends beyond the realm of the service to the public interest into the realm of service to the common goodassociated with a vision of society as a community whose members are joined in a shared pursuit of values and goals that they hold in common, a community comprising persons whose own good is inextricably bound up with the good of the whole, the well-being of the community, and the preservation of its core values (33).

Moreover, the division of labor in multidisciplinary work has its dangers. It can produce people who stay at the level of knowing nothing but their own discipline, missing all sense of belonging to the larger community of humanity. This actually happens throughout our scientific community where the payoffs are for performance in a very narrow field and where there are very few payoffs for larger visions, prophetic dreams and social commitments. How we bring our scientific community to accept responsibility for the consequences of maldevelopment and social inequities is a problem we have not even begun to tackle.

Furthermore, there are very few channels of communication between the multidisciplinary development community and the politically powerful. (Perhaps not so terrible, after all, if our peers are a source of bad advice...). Consequently, the whole apparatus and structure of political power is presently mostly a setting for the making of bad decisions on development issues on the part of the powerful. And the greater the power, the more likely are the decisions going to be bad decisions simply because the organizational structure that surrounds the powerful is designed to prevent them from learning anything and accountability is almost nil, as we saw in the cases of Marcos, Somoza, or Batista. The scholarly community is by no means always right, of course. As we hinted above, it is perfectly capable of giving extremely bad advice to the powerful, especially if they share a conservative (or liberal...?) ideology (18). In short, interest groups at the top (within the bourgeoisie) invariably find the status quo superior to any feasible reform. On the other hand, traditional multidisciplinary teams, or their members, advise these groups at the top seldom taking the consequences of what they have proposed as solutions; it is the poor, the working class who do. Multidisciplinary work, just by being collective has little potential for meaningful positive impact.

In multidisciplinary development work, we too often arbitrarily make some implicit assumptions about distributional issues, i.e., equal access to or distribution of food, or fixed patterns of inequality, taking them as given. These assumptions, plus neglecting the political questions underlying development work, are not a mere oversight, but rather due to a paradigmatic blind spot, ideological in origin. We inherit these blind spots that become an impediment to our more comprehensive understanding of the development process. Therefore, we will have to take some action to shed some light on these blind spots (34, 35).

The problems in the community are there for everyone to see. Yet in our development theorizing we very often appear not to be aware of them or to ignore them. A likely reason for this is that the paradigm(s) of modern development theory have not included the political and distributional factors as relevant to the questions we have asked of reality. The unit of analysis is more often the individual, not the social group or class. Despite so many references to political factors in economic development, most Western development theory is not equipped to integrate such factors as relevant to its purposes.

In liberal political theory, the individual is shorn of his various social and political attributes. Marxist political economy represents a break with that liberal political philosophy. For Marxists, it is not so much the individual who counts, but the group. It is the social class that becomes the main political actor, the historical force. The individual is only a representative of his/her class or else is defined in terms of his/her relationship to the fundamental class struggle of our time (34). These factors, obviously, become central to the purpose of Marxists development theory.

Conversely, in the Western multidisciplinary setting of development theorizing, we still mostly stay within the general corpus of biological determinist thought tending to blur the structural (class-bound) causes or underdevelopment. What we need is actually to tackle, not dodge these central issues. we rather tend to progressively mask basic development issues with technocratic mystifications. It is even more: In multidisciplinary work the interaction between the various disciplinary paradigms actually too often conceals deep causative forces underlying the process of maldevelopment and the design of corrective measures. This uncritical attitude also tends to conceal of take as given the socioeconomic realities that emerge from the growing contradictions of capitalism (20).

THE ROLE OF CONCEPTUAL FRAMEWORKS

The multidisciplinary debate on development, then must not hide (but so far mostly has...) the political aspects that underlie development and the framework in which it is played out. The very concept of development is understood differently depending on whether we look at it from the point of view of the general application of the Western model of development or from a more egalitarian perspective (36). In an effort to come up with an integrated paradigm for development workers, several colleagues have tried to design conceptual frameworks that will avoid any misunderstandings by their peers starling to share the same paradigm. An example of the product of such an effort is the conceptual framework that analyzes the causes of hunger and malnutrition presented in Fig. 1 (37). As can be seen from the diagram, this novel (but not necessarily new) approach introduces three causal levels underlying hunger and malnutrition. The only thing this (or any) integrated development paradigm is actually offering is a fresh look at old contradictions in society and in overall past and present development work. (Insert Fig. 1 around here).

It is worth noting that the acceptance of a framework such as the one in Fig. 1 presupposes not only the sharing of the concept of a new integrated (unified) paradigm, but also some deeper ideological premises. It seems to me that utilizing conceptual frameworks like the one here presented may be an initial step in the right direction - increasing levels of social and political consciousness - especially it they force us to think about the root causes of the problems we see. The risk, though, is that such frameworks may only help create new conceptual categories, with actions flowing from their use not changing commensurately. Moreover, many traditional scientists and organizations already admit they have problems with the issues raised by these conceptual frameworks, e.g., the UN University's Hunger Program had problems in the early 1980's accepting a framework very similar to the one in Fig. 1, a fact that should make us wonder what would happen if one added the political and ideological parameters in even stronger wording into these analyses. Since then, the same Program has managed to totally avoid confronting issues of this kind (13).

Cynics would probably raise the question whether the conceptual framework is not - to begin with - one more example of dressing a political approach (perhaps even a Marxist one...) in sheep's clothes in order to "sell" it to our peers as a new "scientific and apolitical" approach. This again only highlights the blurred border between paradigm and ideology, thus reinforcing my original point. Finally, although I may have led the reader to believe otherwise all the above is not really intended as a blanket denouncement of multidisciplinary work; it just points out that it is not a panacea as a development strategy or an approach to development per-se. Multidisciplinary teams should rather provide the forum for the expression and dialectical interaction of paradigmatic and ideological perspectives and contradictions in development work in an environment that does not intentionally or involuntarily suppress any perspective and that incorporates - de facto - the perspectives of the beneficiaries.

Ideology

Webster's, again, defines ideology as: Visionary theorizing: manner or content of thinking characteristic of an individual or class; intellectual pattern of any culture or movement; integrated assertions, theories and aims constituting a politico-social program. Ideological values and duties that follow from the above definition are imprinted on us through the family, through education and through the social environment we grow up in. These values are thus not universally shared and are closely bound to our social class extraction. Ideology, as a content of thinking and as an intellectual pattern, reflects the involuntary elements of our behavior that are part of our indelible (class) heritage. On the other hand, ideology as an integrated politico-social program, is the result of a voluntary internalization of the values of a given society, whether real or utopic (38). It is ideology that ultimately channels our social behavior in predictable directions.

Ideologies are more deep-rooted than paradigms. You can break a paradigm with a series of new discoveries, but not an ideology. In our case the dominant research "Establishment" assures and promotes its own ideological choices through what it funds, disseminates, and honors (13). Yet people can change their class stand and their ideology. Ideology is not just passive; it is or can be active. Epistemologically then i.e. in terms of the origins of our understanding - and no matter how one looks at it - paradigms are ideologically rooted.

Ideology and legitimacy

As Montessori already understood, societies mold their educational systems to fit the interests and goals of those in control, often with the unfortunate consequence of limiting their people's development. The process of ideological legitimization is the express attempt to put all relevant issues of an epoch within the boundaries of existing social values. It is used to rationalize and justify people's personal and professional goals and interests in society. This legitimacy even allows us to uphold dogmatic traditions with candid ingenuity (39, 40, 41).

All new scientific development thus necessarily involves a phase of adjustment and conflict that forces its actors to justify (legitimate) their pretensions in ideological terms as determined by the existing array of ideological opportunities to pick from. There actually is something like a market of raw materials to fabricate ideologies. A new field (development included...) always arises in ideological conflict with the milieu that will be more or less hostile to it.

The importance of ideological change should, thus, not be minimized because it is affecting us in development work and it is these philosophical and attitudinal changes that will ultimately have to be translated into practical changes to be introduced in our work in social and economic development (42, 43). Where some see only different paradigms in a struggle in development theory there invariably are also ideological barriers or differences at play. The decision to opt for one or another paradigm is not only logical, it is heavily ideological as well. If the needed changes in the optic with which we look al problems eventually occur, it will necessarily lead us to focus on a complete new set of problems, especially when we feel old theoretical models are falling behind the empiric evidence coming in (44).

As long as new theories about underdevelopment and about priorities for the Third World address the task of combating underdevelopment's consequences and not its causes - or combating either causes or consequences within the capitalist-dependent framework of development as we mostly see it today - these theories will remain ideologically biased favoring: status quo. They will thus be subject to growing challenge by the proponents of empowering, indigenous, and self-sustaining development. Any attempt to prove the feasibility of solving the problems of underdevelopment in the Third World within the framework of capitalist alternatives is itself highly ideological because the ideological function of any development work manifests itself primarily through the context in which it is carried out and not through its technical content. Western development, as an instrument of legitimation of the capitalist mode of production, is accountable for perpetuating the system that generates maldevelopment. It is thus the structural and social context of development work that renders it ideological in character. The priorities finally chosen will invariably express class relations and will lead to a certain (foreseeable) distribution of economic and political power. (20).

We, therefore, need to keep in mind that we ultimately relate to our peers and work-counterparts in ideological or sub-ideological terms. We link with the like-minded. Only once we agree on some common ideological bases do the more technical issues (on which we now spend so much time) become relevant and important.

Ethos and norms

The ethos has been defined as a complex of internalized norms and professed values. The scientific ethos, then is a set of norms that regulate the activity of men and women of science. Is this ethos thus a sort of "professional ideology" that justifies our enterprise and legitimizes our pretensions? (45). Ethos is often misused as a synonym of ideology. The ethos has ideological roots, but is not ideology. The ethos expresses itself as imperatives legitimized by the institutional values that mold our scientific conscience. There are thus psychological sanctions for certain ways of thinking and of acting. Moral indignation is directed against the violations and the violators of the ethos (45, 46).

The ethos of science has not been coded, but it can be inferred by the moral consensus of the scientists as a "climate" of the dominant values and feelings. For most of us scientists, the norms of science have not only become moral issues, but are also behind the technical prescriptions we abide by. Actually, the norms we abide by can be inferred from our everyday work behavior. The fact that these norms are often validated and reinforced by the (mostly bourgeois) milieu outside the scientific community, says nothing about their ultimate validity (45, 46). The ethos of science is based on norms that "protect" the method as a set of accepted techniques. These norms, however, are only general and cannot be used expediently when it is necessary to discern between concrete alternative courses of action, i.e., they are only "the terms of an ideology that cannot give immediate recommendations for action". One can, therefore, safely say that consensus in the scientific community has both moral and ideological bases (45,46).

Where do the actual norms of our modern scientific ethos come from, then? Are they external or internal factors to science itself? To what extent are norms effectively internalized during our training or through our daily work? Our training certainly represents a process of socialization in the prevailing "scientific ideology." We learn to think in tune with the predominant scientific paradigm. And this has to do with the textbooks and journals used and with the analysis of concrete solutions given to scientific problems by those books and journals. Whether we like it or not, such an internalization of the norms is quite authoritarian and dogmatic (45, 48). Unfortunately, life is rarely as simple outside the covers of a textbook as inside... To make things more complicated, we development scientists often show de facto ambivalence towards the institutionalized norms of the scientific ethos because, for some among us, the scientific ethos not only legitimizes our day-to-day activity, but through the latter it also imposes on society primarily middle class social values and pretensions (45, 49).

Conflicts in the terminology?

As should be clear by now, the definitions of paradigm and ideology tend to somewhat overlap. This overlap inevitably leads to some confusion in determining what is ideological and what is paradigmatic in a given specific context, or where one stops and the other starts (if both are part of the same continuum...). The latter confusion becomes even greater as colleagues of ours are calling for integrating and unifying development paradigms.

The question that naturally follows, then, is: Can an integrated natural/ social sciences development paradigm be born through multidisciplinary work without bridging the underlying ideological gap (if there is one...) among the practitioners of the contributing disciplines? It would seem that only arriving at a "minimum consensus package" amongst them is not enough; any viable unified approach would seem to require a shared underlying ideology. The dilemma, then, is whether the creation of such a unified paradigm should be a priority goal in development work, as there seems to be a call for. I contend that if this exercise becomes a stepping stone to bridge the ideological gap among the practitioners of development coming from, different paradigms, it becomes a worthwhile exercise. otherwise probably not. Now, how this collective work should be done to arrive at a unified paradigm is not very clear either and, in any case, is not an easy task. Moreover, we can foresee that it will become an almost inevitable source of conflict if the actors are not willing to compromise in order to start sharing some new views on development. (Conflict, under these circumstances, is of course not necessarily bad...).

Because paradigms have very selective "blind spots" - often ideological in nature - w frequently see discrepancies between ends and means pursued by the practitioners of the disciplines related to development work. The Western development paradigm follows that rather rigid slicing or analysis of reality into different disciplines that was mentioned earlier. Underdevelopment however, cannot be compartmentalized if it is to be fully grasped: it is a total situation in which every element plays a part. A more political approach to underdevelopment is needed rather than a less political one.

And this entails a clear-cut choice of ideology with very few (if any) shades of gray to choose from in between. This leads me to reaffirm my belief that paradigms are submanifestations of ideology. As such, the two do not clearly replace each other: rather, they continue to coexist and overlap. The evolution from a disciplinary paradigm to an integrated one - as the one we seem to seek - is seldom smooth; it rather involves a quantumleap with inescapable ideological connotations; (23).

Some common characteristics of our disciplines of origin:

The traditional image of the disciplines we were trained in as being highly successful scientific enterprises is only partially valid. "Normal science" (the practice of "puzzle- solving" in science) is actually a highly directed and regulated activity based on the rigorous previous socialization of its practitioners within the respective disciplinary paradigm(s). We constantly look for the right connection that can relate our problems (in development) with the accepted body of existing scientific knowledge; that is, it is we who are put on trial (can we find the connection?), not the prevailing theory (does that latter apply to what we found?).

Thus it is the past that governs our present activity (forcibly leading us to conservative approaches to development...?) Since we use the rules provided by the paradigm within which we operate, normal science is only "achieving the expected in a new way" (50, 51). All this really shows is that if we are wrong, we are wrong in good company...

In normal science, the main game is to resolve enigmas ("puzzles") and not to challenge or confront theories, especially in development work. Some room needs to be made for intra- and inter-theory disputes to allow for an eventual displacement of one conceptual system by another. A nonconflictive and consensual conception of science reinforces the scientific and the social status-quo. (52, 53).

There is an academic (university-based) monopoly over the definition and the practice of the sciences, but the question is: Is this academic model of the sciences actually in a phase of decline? Is science only one or are there many science? At present, it is almost impossible to consider science as a compact whole. There is a notorious difference between the "old" and the "new" sciences, the "hard" and the "soft" sciences. There is, therefore, probably not a sole or unique ethos of science either, thus adding further complication. Does this ethos of science change under changing historical configurations (47, 54, 55)? If yes, how fast? Have we been flexible enough to change accordingly? If yes again, how far behind are we?

Subjectivity of the sciences

As in most professions, scientists often approach a given problem with some inherent bias (56). What we call "truth" actually depends on what there is and on the contribution of the thinker. There is thus a conceptual human contribution to what we call "truth" (Einstein). Unfortunately, development organizations and practitioners are sometimes more concerned with how things look rather than with how they really are. And when dealing with images rather than with reality, they eventually get strangled in their own lies or half-truths. On the other hand, organizations can also be looked upon as emanations of a hierarchical power structure serving the interests of a particular ideology so that half-truths may be accepted knowingly if they serve a given purpose...

The world, then is constituted by facts outside ourselves and by factors that we introduce in it. The concepts we impose on the world may not be in the best interest of the poor, though, and we had better revise them with the benefit of their input (57). It is ultimately we development workers who contribute to the irrefutability of the track towards maldevelopment Western development is presently following. Development praxis can thus be redirected onto a more effective track, leading to more lasting solutions to poverty and powerlessness. Values are not part of the sciences, but of the ideology of the inquirer. For Gunnar Myrdal, research is always, by logical necessity, based on moral and political values. The inquirer, he prescribes, should get involved in participatory research to show social and political commitment and to achieve what the grassroot groups want, i.e., action that brings about real changes (58).

Rationality and objectivity in the sciences are thus relative. They are possible only if there exists a previous agreement on some fundamentals (that can be paradigmatic and/or ideological) (59). To no one's surprise, then, the sciences can confound factual judgements with value judgements because nothing is learned in abstract: we learn to share a tradition and use value judgements all the time (57).

"Choose a group of persons, the most capable and motivated ones: train them in a given science and its specialties; soak them in the predominant value system or ideology in that discipline and finally let them make a choice." That is the natural history of scientific development as we know it. Whatever the nature of scientific progress, we ought to size it by bringing to the surface and exposing what the group of scientists behind it values, tolerates, despises or feels contempt for (41).

It is normal for us to show a continuous interest in searching for "competent responses" in our respective lines of work, but in so doing we are being ideologically biased. Even false or outmoded theories can be defended for a long time with "progressive", biased jargon. to get at the "why" of facts, we cannot just ask "how" they happen and then merely put these facts in correlation by no-matter-what means (e.g., mathematical or statistical means) to get to the "why". The imponderables of reality surpass our mere computational capacity (44, 60, 61, 62). Competent responses in development will have to come from altogether new, updated theories.

Scientists tend to accept the postulates and results presented by members of their own group of reference easier and without criticism whereas they subject to rigorous criticism the postulates and results coming from an (ideological or otherwise) opposing group. It is typical for the nonradicalized or not-too-critical scientists among us to assimilate the prevailing dogma, not feeling any urge to challenge it. These "routine scientists" in our midst, more than real scientists, have become individuals who know how to apply a technique (59, 63).

7

Sustainable Product Design and Manufacturing

HISTORY OF SUSTAINABLE PRODUCT DESIGN, TOOLS & METHODS

HISTORY OF SUSTAINABLE DESIGN

Eco-design, which attempted to integrate environmental concerns into the design of artefacts and the built world did not emerge as a discrete discipline until the mid 1970's. However, its roots lie in earlier concerns expressed by ecologists such as Leopold (1887-1948) and Naess (1912)); and socio-economic and political critiques and visions for a more sustainable worldview, offered by people such as Mumford (1895-1990); Bookchin (1921)) Packard (1914-1996); Nader (1934) and Schumacher (1911-1977). Many cite the publication of Rachel Carson's Silent Spring in 1962 as a catalyst for environmental concern and Victor Papanek's Design for the Real World in 1971 as sparking the emergence of the eco-design movement.

The Bruntland Report (1987) prompted the second wave of environmental concern along with the green consumer revolution which regarded design as important in the development of more mainstream eco-products and in enhancing consumer acceptance of these. By the early 1990's designers were beginning to take these messages on board stimulated by 'design for the environment' being placed on the corporate agenda (Mackenzie 1997; Burrall 1991).

Throughout the 1990's to the present day, a plethora of exciting new developments are pushing the boundaries of design. Ezio Manzini (2003) is part of a growing movement to broaden the remit of designers towards creating a vision for sustainable everyday life that re-defines the need for different types of goods and services. New approaches based on biological and ecological systems (inspired by the earlier work of D'Arcy Thompson (1921) as well as natural patterns and morphological processes (such as Alexander's Pattern Language 1977) are shifting the agenda. At the same time, the work of Walter Stahel at The Product Life Institute and others are spearheading the new dematerialisation trend via reconfiguring 'product service systems'.

TOOLS AND METHODS

There is a rich and diverse range of tools, methods and principles for Sustainable Product Design. For a comprehensive review see Sherwin (2001) and Charter and Tichner (2001). In this section, we provide illustrative examples of the different types of tools and methods that are available. Most of those surveyed are concerned with improving resource and energy efficiency of single products as well as reducing waste and toxicity impacts (such as Factor 4, 10 and 20 models).

Design methodologies informed by this approach tend to start with a product focus, initially incorporating single issues (such as the use of some new material such as recycled plastic or energy efficiency solutions). More sophisticated models adopt a lifecycle approach, addressing all impacts along a product's lifecycle (material extraction, production, distribution, use and disposal) which in extended versions includes full triple bottom line impacts.

Broader approaches extend beyond single products to address larger ecological and social (eg. relating to consumption and lifestyle) questions in a systemic framework. These methods draw from principles of ecology, biology and a vision of a future sustainable society to inform the different types of products and services which would then be needed.

The tools and methods are presented below in rough order of those that are primarily product focused to those that are user/consumer focused.

Hierarchy of Waste Management: This is an early design model whose primary emphasis is on waste reduction and management. It promotes the 'reduction' of waste as the primary goal, followed by the 'reuse' of product, components and materials in design.

Third on the hierarchy is 'recycling', some way below its perceived priority design status. This model led to the earliest and perhaps most simple eco-design product focused strategies – reduce, reuse, recycle. (Sherwin 2001; 23).

Factor X eco-efficiency concept:

Eco-efficiency is a product based approach that focuses on increased resource and energy efficiency through technological innovation. Von Weizsacker, Lovins and Lovins (1997) have argued that greater eco-efficiency could allow consumption to be doubled while environmental impacts are halved, the 'factor four effect'. Practical examples of products that have achieved this level of increased energy and materials productivity include office furniture by Herman Miller; the 'hypercar' developed by the Rocky

Mountain Institute; and the FRIA Cooling Chamber designed by Ursula Tichner (Cooper 2000).

Life Cycle Assessment (LCA):

Life cycle principle/assessment is a concept in which all the stages of a product 'life' are considered through design (material extraction; production; distribution; use and disposal). It is one of a range of tools that support life cycle management, - minimizing environmental burdens throughout the product/service lifecycle. Life cycle approaches consider local as well as global impacts and attempt to incorporate environmental factors into early stages of design. Decisions made at various points in the lifecycle have effects both up and down stream, for example recycling wastes back into the lifecycle system can potentially reduce resource use at its extraction. Extended versions can incorporate social and economic impacts. LCA is one of the key methods recommended in the Product Sustainability Toolbox developed by the Advisory Committee on Consumer Products and the Environment (ACCPE 2004)

Cradle to cradle: This approach calls for more radical design solutions to bring our economic and social systems into harmony with the wider ecological systems on which they depend. For McDonough and Braungart and (2002) design of products and services needs to enrich the environment rather than deplete it. The language of this agenda is more about inspiration, creativity and innovation – moving away from reducing and minimising destructive impacts of products and services – towards restoration and enrichment.

"eco-efficiency only works to make the old, destructive system a bit less so. In some cases it may be more pernicious, because its workings are more subtle and long term" —McDonough and Braungart (2002; 62)

Cradle to Cradle design advocates 'closed resource loops' in which waste from one industry becomes 'food' or raw materials for another. Products are seen as 'technical' or 'biological' nutrients which must return to a resource pool for further use. Technical nutrients include items such as refrigerators, televisions and other appliances and their materials remain within technical cycles. Biological nutrients (the products or services of organic-based materials) are designed to safely return to organic cycles. Also known as 'zero waste' or 'closed loops'. The Cradle to Cradle approach has been pioneered by designers Bill McDonough and Michael Braungart and used in practice by Gunter Pauli in 'ZERI Clusters'.

Product service systems PSS):

Other design innovations that broaden to include the social dimension have been developed by pioneers such as Walter Stahel (founder of The Product Life Institute in Switzerland). According Walter Stahel: "for an increase in resource productivity well beyond factor 4, or even factor 10, innovative strategies for new solutions are needed, attacking problems on a systems level instead of on a product level". (Stahel 2001). The key idea behind PSS is that consumers do not specifically demand products, but rather are seeking the utility these products and services provide. By using a service to meet some needs rather than a physical product, more needs can be met with lower material and energy requirements.

There are many examples of this approach in business to business services (eg. Xerox – photocopying; Wikhahn – furniture; Interface – flooring; Electrolux – industrial cleaning) and some penetration into consumer markets such as mobile phone services. A number of international programs and networks have been set exploring this approach (such as SUSPRONET and MEPSS). PSS can be used at an extended systems level (eg. to understand service needs of consumers) as well as being more narrowly focused on individual product-service combinations.

Bespoke products-services

Small scale, bespoke products provide an alternative to mass production based on high quality, made to measure products that are durable and timeless.

This is as much about addressing sufficiency of consumption as improving efficiency. Hand crafted products (such as shoes and furniture) are an example of product design that incorporates a large service element; fosters durability through consumer attachment, care and repair as well as reducing resource intensity per worker. A case study of the contribution that custom made products (in this case shoes) can make to sustainable design concludes that:

"With the production of hand-crafted made-to-measure shoes, it can be concluded, strategies of sufficiency (better instead of more) can be carried out. This labour-intensive production method allows for the creation of eco-efficient, decentralised and resource-preserving jobs. At the same time, durability, reparability and a high level of appreciation for the product leads to ecological gains. The customer's involvement cements his or her willingness to us the product for a long time". (Ax 2001)

Four step model of ecodesign innovation

This is a model that straddles both product and systems focused approaches. It was developed by Charter and Chick (1997) as a 4 step model of design innovation: Repair deals with end of pipe solutions. Refine product design through the concept of eco-efficiency. Redesign and Re-think will require significant leaps in design strategy from a more systemic perspective.

"To move beyond 'Re-design to Re-think will requires significant leaps in thinking, driven by the emphasis on creative problem-solving and opportunity seeking. An essential element of this process will be the development of a more systematic infrastructure to enable the cyclical flow of resources and energy within the product systems, as outlined in the emerging concept of 'industrial ecology"(Charter and Chick 1997)

Biothinking/Biomimicry: This is a growing field in which the design of products and services mimic nature by utilising biological and ecological principles and processes. Ed Datchefski and Janine Benyus are pioneers in the field in which observations of natural systems have inspired a number of exciting new product designs. The principles they use to guide product design are summarised below.

Ed Datchefski (2001)	Janine Benyus (1997)
Cyclic: products should be part of natural cycles, made of grown material which can be composted or become part of a man-made cycle like closed loop recycling Solar: all energy used to make or run the product should be from renewable energy in all its varied forms, most of which are driven by the sun Efficient: increasing efficiency of materials and energy use means less environmental damage. Products can be designed to use 1/10th of what they did before Safe: products and their by products should not contain hazardous materials Social: product manufacture cannot exploit workers	1. Use waste as a resource 2. Diversity and cooperate to fully use the habitat 3. Gather and use energy efficiently 4. Optimise rather than maximise 5. Use materials sparingly 6. Don't foul your own nest 7. Don't draw down resources 8. Remain in balance with the biosphere 9. Run on information 10. Shop locally

This type of approach shifts design from a largely technological and industrial to an ecological and biological paradigm. This approach can be used for product focused design (eg. using functional solutions informed by biological organisms to inform product function and aesthetic) as well as extending the principles to wider systems (eg. communities, organisations, local economies)

USER-CENTRED DESIGN

There are several collaborative processes emerging in the area of sustainable design that seek to involve both practitioners and users in the research and design phases of products and processes. This approach can also be seen as a socially inclusive process that strives for consensus through design and which can successfully include practitioners from a wide variety of disciplines and knowledge areas. This approach is characterised by a holistic, system-wide view of design problems and their contexts. It has been pioneered by Christopher Day in architecture and by Hilary Cottam and colleagues at the Design Council in the design of public services and buildings (in health, education and prisons).

Pattern Language

This design method developed by Christopher Alexander (1977; 2002) and colleagues at The Centre for Environmental Structure in Berkley starts from a systemic perspective which sees relationship, pattern and context as the primary starting point for design. The design method involves deep observation of social and natural patterns as the basis for creating harmonious and coherent relationship between the human-made and the natural environment. Natural pattern has provided the inspiration for the design of the new education centre at the Eden Project. The entire building is based on the Fibonacci sequence and the golden proportion which form the basis of the geometry which underlies plant growth.

Manzini's principles

Manzini has developed some of the most advanced thinking in the practice and theory of ecodesign that starts from a systems wide perspective and questions the role that goods and services play in a sustainable future. He uses three principles (Manzini 1993):

- From consumption to care: Developing products that require care and with which the use can establish an emotional relationship
- From consumption of products to utilisation of services: Looking at the concept of utilisation, going beyond the notion of possession and personal consumption
- From consumption to non-consumption: In which the reduction of needs can experienced as an increase in social quality.

Ezio Manzini in Sustainable Everyday, (2003) starts from a wider system perspective to envision sustainable everyday life in an urban context. This then creates a challenge for design in re-defining the need for different types of products and services:

"...the transition towards sustainability in its everyday dimension, can be described as follows: in a short period billions of people must redefine their life projects. Although differing greatly, the new directions they can and will want to take have a common vector—one which should take us in all our diversity toward a sustainable future".

These latter design approaches tend to start from wider systemic principles and then work backwards to re-define what types of products and services would be needed and how they would be delivered in a sustainable society. Starting from this perspective leads to more innovative and radical transformation of product and service design.

THE BUSINESS RESPONSE TO THE CHALLENGE OF SUSTAINABLE DESIGN

Our research shows some interesting generic characteristics of how business has responded to the sustainable design challenge. Overall there has been a general shift in focus of 'the product development process' from end-of-pipe solutions to designing-out and front-end solutions further upstream. This has seen the emphasis change from tackling single issues to considering the full impacts of the products or services and can be described as a move to the earlier product development stages.

The approach has been incremental (i.e. small step changes to existing products) and chiefly technology-driven using lifecycle thinking and approaches – rather than a more systemic or radical approach to devising new business models and new types of products and services. There also seems to be a bias towards measurement and assessment of product impacts, rather than product improvement – with an apparent assumption that the former automatically leads to the latter.

The gap between the number of product impact studies & tools versus the number of improved or sustainable products would confirm this. In the sectors reviewed for this report, government legislation (such as producer responsibility regulations) is unlikely to provide sufficient incentives for re-design though it can and does help improve the impact of products in other 'non-designed' ways.

In general, the practices of small and medium size enterprises (SME's) are not seen as being as advanced in sustainable design methods as multinational business though it remains to be seen if corporate products are actually more sustainable.

CORPORATE DRIVERS AND BARRIERS TOWARDS SUSTAINABLE PRODUCTS AND SERVICES

Corporate Drivers	Corporate Barriers
Policy, regulation and voluntary standards (at sectoral, national and EU level) Cost savings and efficiency ('eco-efficiency' gains) Labelling – with mixed success Public, NGO and media pressure – acts as both motivator and barrier Marketing and consumer driven – 12% rise in demand in 2004 Brand value/reputation – strong links between sustainable design and branding Corporate leadership – Many business programs & policies - as well as emerging corporate leaders - are now mandating sustainable products and design.	Business case – developing a robust business case that reaches beyond short term single issues that vie for attention Lack of demand – growing markets but still niche/small in terms of market share Internalising external costs – some efficiency gains but often results in net business costs Lack of policy incentives – insufficient tax breaks or grants

The role of the designer: Some interesting themes are drawn for the role of the designer in the sectors (Fast Moving Consumer Goods & Electronics) reviewed for this report:

Driving sustainable design:

- Sustainable products are only one (small but growing) part of a broader sustainability agenda within business usually driven by sector specific issues (i.e energy use/recyclability in electronics; ingredients & formulation in FMCG's)

- Sustainable design seems to be driven and work successfully in business through: hook up to a corporate/business program or policy; when there are pressing SD issues for the sector/business (i.e. water scarcity); when there was a sector or business scandal
- Sustainable design – or ecodesign/design for the environment as it is often known in business – is an increasingly accepted and practiced term. It helps connect a sustainability strategy to products and materialise a management system.

Influencing sustainable product/design strategy:

- Sustainability issues are best considered upstream in the business and product development process. Often, designers have little early strategic influence here.
- Industrial designers are only part of the design process – others marketers; engineers/developers; strategic functions; and/or research and development play a larger role and have more significant influence over the resultant product impacts.

Practices of sustainable design:

- There is evidence that the development of sustainable products can (and at present is) happening without designers and is taking place in other 'product development' departments (such as mechanical, electrical or chemical engineering). In certain cases design is externally outsourced, meaning it is seen and managed as an external service provider
- At present there seems little link between corporate sustainability generally and even sustainable product innovation targets specifically – and design skills. Companies often use other non-design disciplines to conduct this
- Current business models and approaches view sustainability as a largely technical/technological issue which accounts for the predominant engineering approach

Moving forward

- There are one or two weak signals from companies that show possible paths as well as a unique set of skills for designers to contribute to sustainability:
  - The emergence of a non-technical, non-material approach to sustainable design – especially relevant for creative designers - that is more consumer (and sustainable consumption)-focussed and has the potential to challenge consumer mind-sets and

purchasing patterns. This might see more emphasis user centred and communication design development.

- Occasionally companies use design skills for fresh thinking and new insight - to conceive new ideas and approaches and shed new light on sustainability problems as well as future visioning.

Transferable and identified design skills could make a rich contribution to sustainability within business beyond the approaches to sustainable product design. We need to further define this contribution and make the business case.

INTEGRATING SUSTAINABILITY IN THE PRODUCT DEVELOPMENT PROCESS

One of the challenges in mainstreaming sustainable product design in the UK is to identify who is involved in the product design/development process. Both MPDs and SPDs agree that professional designers often enter the process late and therefore contribute little more than styling, 'add on' functionality or efficiency. In contrast, research shows that the most significant social/environmental impacts are achieved when sustainable design is incorporated at the earliest/conceptual stages.

Therefore, it is important to recognize that design is an activity many stakeholders take part in; engineers, project directors, marketing/brand experts, business strategists and manufacturing/production, users/consumers to name a few.

Often decisions about whether to incorporate sustainability concerns into the design and development of a product happens at strategic corporate levels at which designers rarely have access. This may account, in part, for their general lack of awareness/understanding of sustainability as it applies to government policies. Most designers also fail to see an opportunity to create new pathways for consumers with respect to sustainable products and the shift in consumption patterns and lifestyles that they can stimulate.

Challenges/barriers to designing sustainably

Both MPDs and SPDs agree there are many products on the market that display aspects of sustainability but generally feel that 'best practice sustainable product design' has yet to be achieved for a variety of reasons: failure of business to place sustainability on the design brief (designers believe business isn't asking because consumers aren't demanding/ government isn't requiring), lack of awareness (on the part of designers, business and consumers), lack of appropriate tools/methods/knowledge sharing and the prohibitively large skill set that SPD requires.

MPDs are especially challenged to acquire the necessary skills and knowledge required to design sustainably, with many saying they are at a loss to know where to begin. They are challenged not only to acquire the necessary design skills but to understand what questions to ask at the critical early stages of a project. SPDs in contrast, have the skills and knowledge but often have trouble convincing clients to place sustainability high on their list of concerns. All designers queried felt there was tremendous potential for government to motivate business but felt it should be doing more.

This reflects the generally passive role that designers take with respect to business. MPDs especially stated that they were unlikely to begin to design sustainably until their clients asked them to do so (and noted that this would not happen until consumer demand increased). A few SPDs are aware of the new raft of government policies and see them as opportunities to define not only new markets but new business models, (such as designing sustainable products and licensing the design to manufacturers) but they are in the minority.

Another concern held by both MPDs and SPDs is the large and varied skill set required to design sustainably. Over 30 different skills have been identified in this study which range from the highly specific (prototyping, materials selection) to the high-level/multidisciplinary (ability to facilitate change, strong collaborative skills and an understanding of global issues/trends). Most SPDs felt that on the job training was essential to prepare young designers for this area of specialty and many MPDs are simply not interested in taking on an intense course of study/skills acquisition. The following motivators and barriers to sustainable design were cited by MPDs and SPDs:

Motivators	**Barriers**
Competitive/economic advantage	Requires larger skill set
Desire not to be 'left out'	Designers not in influential positions
Compliance w/gov. regulations	Unpopular/misunderstood
Market/consumer demand	'Tough sell' to consumers/clients
Corporate responsibility concerns	Perception of higher cost w/SPD
Personal motivators	Lack of appropriate tools/methods
Connection to nature	Lack of government support
Alignment of ethics with profession	Lack of consumer demand

Motivating Designers

When designers were asked how to forward sustainable product design in the UK, both groups responded that public awareness of sustainability was key (in stimulating consumers to ask for sustainable products and for business to place it high on their agenda). Without acquiring additional business/entrepreneurial skills, most designers are unlikely to take the lead in identifying opportunities that government policy may provide.

Designers also feel that creating more/better case studies and design competitions (with rigorous guidelines) that were highly publicized and supported by government is a way to build interest and support from within the design community. This underscores the relatively passive role that designers tend to take with regard to sustainable design and their contention that they must wait for consumers to demand sustainable products from business who in turn will require them to design sustainably.

The challenge

The challenge is how to raise awareness of the business and creative opportunities that sustainable product design holds for designers. This will take a variety of forms; new markets and business models stimulated by government legislation; the potential for competitive edge that sustainable products can represent for business; and new creative challenges/modes of designing.

This will require support/funding at the government level (perhaps led by the Design Council) to raise awareness of these issues with designers as well as new formats/methods for ongoing and life-long education to provide them with the necessary skills. On the positive side, both MPDs and SPDs see sustainable design as essential and inevitable and envision a future in which 'sustainable design' is not a specialist area of design, but rather an attribute of good design.

NEW CHALLENGES AND OPPORTUNITIES FOR WORK ORGANIZATION

Work and the workplace are essential elements of industrial and industrializing economies. Work is combined with physical and natural capital to produce goods and services.

The workplace is the place where the comparative advantages of workers and owners/managers create a market for exchange of talents and assets. Beyond markets, work provides both a means of engagement of people in the society, and an important social environment and mechanism for enhancing self-esteem.

Finally, work is the main means of distributing wealth and generating purchasing power in dynamic national economic systems. This essay explores the complex relationship between employment, and the increasingly unsustainable and globalizing economy; the changing nature of industrial economies presents new challenges and opportunities for the organization of work in both industrialized and industrializing countries.

THE UNSUSTAINABLE INDUSTRIAL STATE

Those that argue that the industrialized state – whether developed or developing – is currently unsustainable emphasize a number of problems. These are depicted schematically in Figure.

The 'environmental problems' include toxic pollution, climate change, resource depletion, and problems related to the loss of biodiversity and ecosystem integrity. The environmental burdens are felt unequally within nations, between nations, and between generations, giving rise to international, intra-national, and intergenerational equity concerns that are often expressed as 'environmental injustice'. The Brundtland formulation of sustainability seems to focus concern on intergenerational equity, but all three kinds of mal-distributions are important.

The environmental problems stem from the activities concerned with agriculture, manufacturing, extraction, transportation, housing, energy, and services -- all driven by the demand of consumers, commercial entities, and government. But in addition, there are effects of these activities on the amount, security, and skill of employment, the nature and conditions of work, and purchasing power associated with wages. An increasing concern is economic inequity stemming from inadequate and unequal purchasing power within and between nations – and for the workers and citizens of the future.

Whether solutions involving industry initiatives, government intervention, stakeholder involvement, and financing can resolve these unsustainability problems depends on correcting a number of fundamental flaws in the characteristics of the industrial state: (1) fragmentation of the knowledge base leading to myopic understanding of fundamental problems and the resulting fashioning of single-purpose or narrowly-fashioned solutions by technical and political decision-makers, (2) the inequality of access to economic and political power, (3) the tendency towards 'gerontocracy' – governance of industrial systems by old ideas, (4) the failure of markets both to correctly price the adverse consequences of industrial activity and (5) to deal sensibly with effects which span long time horizons for which pricing and markets are inherently incapable of solving. The solutions to these system problems will be explored in the context of their implications for the organization of work.

Globalization

'Globalization' has at least three distinct meanings [Gordon, 1995], with different implications for workers and working life. 'Internationalization' is the expansion of product/service markets abroad, facilitated by information and communication technology (ICT) and e-commerce, with the locus of production remaining within the parent country. 'Multi-nationalization' is where a (multi-national) company establishes production/service facilities abroad, to be nearer to foreign markets and/or to take advantage of more industry-friendly labour, environmental, and tax policies, while maintaining research-and-development (R&D) and innovation-centered activities in the parent country.

The third meaning is the creation of strategic alliances, what some call 'transnationalization,' in which two different foreign enterprises merge/share their R&D and other capabilities to create a new entity or product line. Those concerned with enhancing trade are especially worried about barriers to internationalization, while those concerned with possible erosion of labour/environmental standards bemoan the consequences of multinationalization. Transnationalization may lead to industrial restructuring with unpredictable consequences for national economies. All three kinds of globalization raise questions of excessive market, and hence political power where concerns for profits overwhelm democratic and ethical values.

Globalization raises new challenges for governance, especially vis-à-vis the roles of government, workers, and citizens in the new economic order. Within nation-states, the extent to which the 'externalities' of production – adverse health, safety, and environmental effects – are internalized differ according to the differential success of regulation/compensation regimes and the extent to which economies incorporate the ethics of fair play in their practices.

There has been a constant struggle to establish good labour and environmental standards/practices within nations. With the advent of globalized, competition-driven markets, attention has now shifted to the harmonization of standards through ILO conventions and multi-lateral environmental agreements, with only a modicum of success.

Countries are slow to give up national autonomy, and only where there is a trend toward significant economic integration (as in the EU) are there successes at harmonization. But globalization has brought an even more complex set of challenges through the creation of trade regimes – such as the WTO, ASEAN, and NAFTA – where the term 'fair trade' means the elimination (or equalization) of tariffs and so-called non-tariff trade barriers, which place labour and environmental standards at odds with trade objectives.

The trade regimes promote international laissez-faire commerce; and rights-based law/protections and market economics have become competing paradigms for public policy and governance. Government plays very different roles when is acts as a facilitator or arbitrator to resolve competing interests, than when it acts as a trustee of worker and citizen interests to ensure a fair outcome of industrial transformations [Ashford, 2002]. The differences are pronounced when stakeholders have largely disparate power – or when some are not represented in the political process, as in the case of emerging or new technology-based firms.

John Rawls argues that no transformation in a society should occur unless those that are worse off are made relatively better off [Rawls, 1971]. Operationalizing a Rawlsian world has its difficulties, but law operates to create certain essential rights that enable just and sustainable transformations. These include the right-to-know, the right to participate in decisions affecting one's working/non-working life, and the right to benefit from transformation of the state or global economy. Struggles won at the national level are now being eroded by a shift in the locus of commerce. Without consensus about fair play and the trustee institutions to ensure fair distributions from, and practices in, the new global economy, equity and justice cannot be achieved. It is now agreed that future development must be 'sustainable,' but that means different things to different commentators.

Sustainable development must be seen as a broad concept, incorporating concerns for the economy, the environment, and employment. All three are driven/affected by both technological innovation [Schumpeter, 1939] and by globalized trade [Ekins et al., 1994; Divan and Walton, 1997]. They are also in a fragile balance, are inter-related, and need to be addressed together in a coherent and mutually reinforcing way [Ashford 2001]. Technological innovation and trade drive national economies in different ways [Charles and Lehner, 1998].

The former exploits a nation's innovative potential, the latter its excess production capacity. Innovation-based performance is enhanced by technological innovation and changing product markets, characterized by fluid, competitive production. Cost-reduction strategies are enhanced by increased scales of production and/or automation, usually characterized by rigid, mature monopolistic production.

Economies seeking to exploit new international markets may enjoy short-term benefits from revenues gained as a result of production using existing excess capacity, but they may ultimately find themselves behind the technological curve. Performance-driven markets may be slower to gain profits, but may outlast markets driven by cost-reduction strategies. The consequences for workers may differ as well.

Increasing labour productivity, defined as output per unit of labour input, is a concern in nations pursuing either strategy. But labour productivity can be improved in different ways: (1) by utilizing better tools, hardware, software, and manufacturing systems, (2) by increasing workers' skills, and (3) by a better matching of labour with physical and natural capital and with information and communication technologies (ICT).

Theoretically, increasing worker productivity lowers the costs of goods and services, thereby lowering prices -- and ultimately increasing the demand and sale of goods and services. Depending on the markets, it can be argued that more workers may be subsequently hired, than displaced as a consequence of needing fewer worker to produce a given quantity of goods and services.

This optimistic scenario assumes a continual throughput society with increasing consumption. However, the drive toward increased consumption may have dire consequences for the environment [Daly, 1991]. In addition, questions arise as to whether, in practice, (1) labour is valued, and paid, more or less after productivity improvements, (2) there are positive or negative effects on job tenure and security, and (3) more workers are hired than displaced. The answers depend on the sources of the increases in worker productivity and the basis of a nation's competitiveness.

Innovation-based performance competitiveness presents opportunities for skill enhancement and building optimal human-technology interfaces, while cost-reduction strategies focus on lean production (with worker displacement), flexible labour markets, and knowledge increasingly embodied in hardware and software rather than in human capital. The consequences for workers are different for these two strategies. The former strategy rewards and encourages skill acquisition for many, with appropriate financial benefits for those workers.

The latter creates a division between workers, some of whom are necessarily upskilled and many whose job content is reduced. Different national strategies might be pursued, reflecting different domestic preferences and culture, but there are further implications, depending on the extent to which trade drives the economy. Interestingly, the US is globalizing and focusing on expanding markets abroad, while the EU is selling a smaller amount and percentage of goods and services outside its borders, focusing instead on integrating its internal markets in which its various members compete on performance [Kleinknecht and ter Wengel, 1998]. In the US, wage disparities are large and increasing, while in some parts of the EU – notably the Netherlands – wage disparities are much smaller and decreasing.

The changing global economy, however, presents challenges for all nations as concerns for the number of jobs, job security, wages, and occupational health and safety increase. In the private sector, labour needs a role in choosing and implementing information-based technologies; in the public sector there is a need for integrating industrial development policies with those of employment, occupational health and safety, and environment. From the perspective of labour, these require implementation of the right to know, the right to participate, and the right to benefit from industrial transformations.

The right to know has been described elsewhere [Ashford and Caldart, 1996: Chapter 7] and includes the workers' right to know/have access to, and the employer's/manufacturer's corresponding duty to inform/warn workers about scientific, technological, and legal information. Scientific information includes chemical or physical hazard/risk information related to product or material ingredients, exposure, health effects, and individual or group susceptibility [Ashford et al., 1990].

As important as information about hazards is, information about technology is the key to workers being able play a role in reducing risks. This kind of information includes not only knowledge about pollution/accident control and prevention technology, but also technology options for industrial production. Knowing how production might be changed to make it inherently cleaner and safer, and the source of more rewarding, meaningful work, is a sine qua non of being able to participate meaningfully in firm-based decisions. Finally, information about legal rights and obligations is crucial for using legal and political avenues for workplace improvement and redress from harm. The right to know is made operational through the right of workers to participate in (1) the technology choices of the firm (through technology bargaining and system design) [Ashford and Ayers, 1987], (2) firm-based training, education, and skill enhancement, (3) national and international labour market policies, and (4) in the setting of national and international labour standards.

While national unions enable workers to work with employers through industrial relations systems, and ILO utilizes a tripartite system that includes labour, management and government, the trade regimes mentioned earlier give little or no participatory rights to labour (or environmentalists) in global economic activities which have potentially significant effects on wages and working conditions.

As trade becomes an important part of national economies, this omission needs to be corrected [European Commission, 2001]. Ironically, under the WTO trade rules, importing countries can restrict imports or place countervailing duties on items that harm their environment, but there is no 'equalizing action' that can be taken if the exporting countries produce those goods unsafely or with adverse environmental effects within their own borders. This reinforces non-enactment or non-enforcement of national health, safety, or environmental laws in the exporting countries, to the detriment of their own workers and citizens. Further, countries may be reluctant to ratify or adopt international accords – including ILO or multi-lateral environmental agreements – in hopes of maintaining or gaining short-term competitive advantage.

Finally, and at the core of justice in the global work life, is the right of working people to benefit from industrial transformations. The right to know and right to participate are essential, but the ultimate rights are those of a fair division of the fruits of the industrial or industrializing state -- and a safe and healthful workplace.

This translates into sufficient job opportunities, job security, and purchasing power, as well as rewarding, meaningful, and safe employment. This can not be left to chance or serendipitous job creation. In formulating policies for environmental sustainability, economic growth and environmental quality are simultaneously optimized, rather than having environmental interventions occur after harmful technologies are in place.

Instead, we seek to design and implement cleaner and inherently safer production. Employment concerns deserves no less a place in center stage; growth, environment and employment must be co-optimized. Systemic changes must be pursued and selected that intentionally benefit employment. Even with better prospects for employment, in an industrial system that continues to replace labour with physical capital, increasing worker capital ownership and access to credit [Ashford, 1998] that turns workers into owners may be an additional necessary long-term option if disparities of wealth and income prevail.

CONCEPTUALIZATIONS OF SUSTAINABLE DEVELOPMENT

It makes quite a difference whether you look at sustainable development as just an environmental issue, or alternatively as a multidimensional challenge in the three dimensions: economic, environmental, and social. We argue that competitiveness, environment, and employment are the operationally-important dimensions of sustainability – and these three dimensions together drive sustainable development along different pathways and go to different places than environmentally-driven concerns alone, which may otherwise require tradeoffs, for example, between environmental improvements and jobs.

A sustainable development agenda is, almost by definition, one of systems change. This is not to be confused with an environmental policy agenda, which is – or should be – explicitly effect-based, and derived from that, a program of policies and legislation directed towards environmental improvements, relying on specific goals and conditions. The sustainable development policy agenda focuses at least on processes (e.g., related to extraction, manufacturing, transport, agriculture, energy, construction, etc.), and may extend to more cross cutting technological and social systems changes.

Table. Comparison of Current and Sustainable Policy Agendas

AGENDA	Competitiveness	Environment	Employment
Current	Improve Performance/Cut Costs	Control pollution/make simple substitutions or changes Conserve energy and resources	Ensure supply of adequately trained people; dialogue with workers Provide safe workplaces
Sustainable	Change nature of meeting market needs through radical or disrupting innovation (a systems change)	Prevent pollution through system changes Change resource and energy dependence	Radical improvement in human-technology interfaces (a systems change)

That current strategy agendas, even those that go beyond environmental goals, are defined as those that are focused on those policies that (1) improve profit and market share by improving performance in current technologies or cutting costs, (2) controlling pollution/making simple substitutions and changes, and conserving energy and resources, and (3) ensuring an adequate supply of appropriately skilled labour, dialogue with workers, and providing safe and healthy workplaces.

We would describe these strategies as 'reactive' vis-à-vis technological change, rather than proactive. They are usually pursued separately and by different sets of government ministries and private-sector stakeholders. At best, policies affecting competitiveness, environment, and employment are coordinated, but not integrated.

In contrast, sustainable agendas are those policies that are focused on technological changes that alter the ways goods and services are provided, the prevention of pollution and the decreased use of energy and resources through more far-reaching system changes, and the development of novel socio-technical systems -- involving both technological and organizational elements -- that enhance the many dimensions of 'meaningful employment' through the integration, rather than coordination, of policy design and implementation.

The kind of innovation likely to be managed successfully by industrial corporations is relevant to the differences between current and sustainable technology agendas. We argue that the needed major product, process, and system transformations may be beyond those that the dominant industries and firms are capable of developing easily, at least by themselves. Further, industry and other sectors may not have the intellectual capacity and trained human resources to do what is necessary.

This argument is centered on the idea of 'the winds of creative destruction' developed by Joseph Schumpeter [Schumpeter, 1939] in explaining technological advance. The distinction between incremental and radical innovations – be they technological, organizational, institutional, or social – is not simply line drawing along points on a continuum. Incremental innovation generally involves continuous improvements, while radical innovations are discontinuous [Freeman, 1992] possibly involving displacement of dominant firms, institutions, and ideas, rather than evolutionary transformations.

In semantic contrast, Christensen [Christensen, 2000] distinguishes continuous improvements as 'sustaining innovation' and uses the term 'disrupting innovation' rather than radical innovation, arguing that both sustaining and disrupting innovations can be either incremental or radical, where the term 'radical' is reserved for the rapid or significant performance changes within a particular technological trajectory.

Thus, in Christensen's terminology, radical sustaining innovation is a major change in technology along the lines that technology has been changing historically, for example a much more efficient air pollution scrubber -- and is often pioneered by incumbent firms. Major innovation that represents an entirely new approach, even if it synthesizes previously invented artifacts, is termed 'disrupting;' and in product markets, it almost always is developed by firms not in the prior markets or business. This is consistent with the important role of 'outsiders' – both to existing firms and as new competitors -- in bringing forth new concepts and ideas [van de Poel, 2000].

Counting only or mainly on existing industries, or on traditionally-trained technical expertise, for a sustainable transformation ignores increasing evidence that it is not just willingness and opportunity/motivation that is required for needed change, but that a third crucial condition -- the ability or capacity of firms and people to change -- is essential [Ashford, 2000]. In some situations they may do so because society or market demand sends a strong signal, but not in all or even in most of the cases.

We argue here that the same holds true for government and societal institutions faced by the triple challenge emanating from new demands in the areas of competitiveness, environment and employment. Intelligent government policy is an essential part of encouraging appropriate responses of the system under challenge, and of assisting in educational transformations as well.

An essential concept in fostering innovative technical responses is that of 'design space.' As originally introduced by Tom Allen et al. of MIT, design space is a cognitive concept that refers to the dimensions along which the designers of technical systems concern themselves [Allen et al., 1978]. Especially in industrial organizations that limit themselves to current or traditional strategies or agendas, there is a one-sided utilization of the available design space. Solutions to design problems are only sought along traditional engineering lines. In many cases unconventional solutions – which may or may not be hi-tech -- are ignored. For that reason radical, disrupting innovations are often produced by industry mavericks, or as a result of some disruptive outside influence (such as significantly new or more stringent environmental regulation and foreign competition, or influence of an outsider to the organization).

The Role of Government

Government is essential for achieving the kinds of industrial transformations that are desirable from an economic perspective, but that are also fair and just in their production and deliverance of goods and services. Among the suggested general functions of government are:

- To provide the necessary physical/legal infrastructure
- To support basic education and skills acquisition
- To invest in path-breaking science and technology development – for enhancing competitiveness, environmental improvement, and job design
- To act as a facilitator or arbitrator of competing stakeholder interests to ensure a fair process
- To act as a trustee of (under-represented) worker and citizen interests to ensure a fair outcome
- To act as a trustee of new technologies
- To act as a force to integrate, not just coordinate policies

More specifically, depending on the specific transformation desired, there is a role for government from the direct support of R&D and incentives for innovation through appropriate tax treatment of investment; to the creation and dissemination of knowledge through experimentation and demonstration projects; to the creation of markets through government purchasing; to the removal of perverse incentives of regulations in some instances and the deliberate design and use of regulation to stimulate change in others; to the training of owners, workers, and entrepreneurs, and educating consumers. The role of government should be considered beyond simply creating a favourable climate for investment. While it is true that 'the government may not be competent to choose winners,' it can create winning forces, and provide an enabling and facilitating role by creating visions for sustainable transformations.

There is continuing debate about the appropriate role of government in encouraging industrial transformations [Ashford, 2000]. Major differences revolve about two competing philosophical traditions: the dominance of unfettered market approaches and a more interventionist, directive role for government through laws and regulation.

Market approaches concentrate on 'getting the prices right,' ensuring competition in capital and labour markets, and increasing demand for a clean environment, product safety, and good working conditions through the providing of information and education.

In contrast, government intervention approaches focus on establishing minimum environmental, product safety, and labour standards and practices; requiring full disclosure by employers and producers of information needed by consumers, citizens, and workers to make informed choices and demands; encouraging technology development, transfer, and infrastructure through a deliberate 'industrial policy;' and requiring decision-bargaining in industrial relations.

Alternative roles for government in promoting sustainable development accomplish different things:

- Correcting market failures by regulating pollution, and by addressing inadequate prices, monopoly power, uncompetitive labour markets, and lack of information achieves static efficiency through better working markets,
- Acting as a mediator or facilitator of environmental and labour disputes/conflicts among the stakeholders achieves static efficiency through reducing transaction costs,
- Facilitating an industrial transformation by encouraging organizational learning and pollution prevention leading to win-win outcomes (based on the concepts of 'ecological modernization' [Jänicke and Jacobs, 2002; Mol, 2001] or 'reflexive law' [Teubner, 1983]) relies on rational choice and evolutionary change that moves towards a more dynamic efficiency, usually over many decades,

- Moving beyond markets and acting as a trustee for minority interests, subsequent generations, and new technologies by forcing and encouraging innovation, through coordinated regulatory, industrial, employment & trade policy transcends markets, moving towards dynamic efficiency within a shorter time horizon.

Conclusion

Recalling that a sustainable future requires technological, organizational, institutional, and, social changes, it is likely that an evolutionary pathway is insufficient for achieving factor ten or greater improvements in eco and energy efficiency and reductions in the production and use of, and exposure to, toxic substances. Nor are fundamental changes in the organization of work likely to emerge through evolutionary change. Such improvements require more systemic, multidimensional, and disruptive changes. We have already asserted that the capacity to change can be the limiting factor -- this is often a crucial missing factor in optimistic scenarios.

Such significant industrial transformations occur less often from dominant technology firms, or in the case of unsustainable practices, problem firms' capacity enhancing strategies, than from new firms that displace existing products, processes and technologies.

This can be seen in examples of significant technological innovations over the last fifty years including transistors, computers, and PCB replacements. Successful management of disruptive product innovation requires initiatives from outsiders to produce the expansion of the design space that limits the dominant technology firms [van de Poel, 2000]. Especially in sectors with an important public or collective involvement like construction and agriculture, this means that intelligent government policies are required to bring about necessary change.

Rigid industries whose processes have remained stagnant also face considerable difficulties in becoming significantly more sustainable. Shifts from products to 'product services' rely on changes in the use, location, and ownership of products in which mature product manufacturers may participate, but this requires significant changes involving managerial, institutional, organizational, and social (customer) innovations. Changes in socio-technical 'systems', such as transportation or agriculture are even more difficult.

This suggests that the creative use of government intervention is a more promising strategic approach for achieving sustainable industrial transformations, than the reliance of the more neo-liberal policies relying on firms' more short-term economic self interest.

This is not to say that enhanced analytic and technical capabilities on the part of firms; cooperative efforts and improved communication with suppliers, customers, workers, other industries, and environmental/consumer/

community groups are not valuable adjuncts in the transformation process. But in most cases these means and strategies are unlikely to be sufficient by themselves for significant transformations, and they will not work without clear mandated targets to enhance the triple goals of competitiveness, environmental quality, and enhancement of employment/labour concerns.

The history of innovation has amply demonstrated that disruptive innovations are feasible, and they may bring substantial payoffs in terms of triple sustainability. They are within the available, but unused design space. However, the general political environment, governmental dedication, and incentive structure have to be right for the needed changes to occur.

Government has a significant role to play, but the government can not simply serve as a referee or arbiter of existing competing interests, because neither future generations nor future technologies are adequately represented by the existing stakeholders.

Government should work with stakeholders to define far-future targets – but without allowing the agenda to be captured by the incumbents -- and then use its position as trustee to represent the future generations and the future technologies to 'backcast' what specific policies are necessary to produce the required technical, organizational, and social transformations. This backcasting will have to be of a next-generation variety of backcasting. It has to go beyond its historical focus on coordinating public and private sector policies. It must be multidimensional and directly address the present fragmentation of governmental functions – not only at the national level, but also between EU, national, regional, and local governmental entities.

There is a great deal of ontology, serendipity, and uncertainty in the transformation process, and the long-term prospects may be not be sufficiently definable to suggest obvious pathways or trajectories for the needed transformations.

Thus, it may be unreasonable to expect that government can play too definitive a 'futures making' role. What follows from this is that rather than attempting tight management of the pathways for the transformations that are sustainable in the broad sense in which we define it in this work, the government role might be better conceived as one of 'enabling' or 'facilitating' change, while at the some time lending visionary leadership for co-optimizing competitiveness, environment, and employment. This means that the various policies must be mutually reinforcing. This newly-conceptualized leadership role – focused on 'opening up the problem space of the engineer/designer' -- is likely to require participation of more than one ministry. Increasingly, ministries of commerce/economic affairs and ministries of environment are working together to fashion a vision of environmental sustainability. What has been missing is a similar proactive role of ministries of labour to interface and integrate employment-related policies into the national and global policy agendas.

THE IMPORTANCE OF TECHNOLOGY, PAST AND FUTURE, FOR SUSTAINABILITY

TECHNOLOGICAL INNOVATION AND TECHNOLOGY CLUSTERS

As mentioned above, technological innovation creating "winds of creative destruction" is widely accepted as the driving force of economic growth in industrialised societies (Schumpeter 1939), historically leading to impressive increases in the standard of living for all citizens of those nations. It is credited as the factor that moves nations from static economic efficiency to dynamic efficiency -- and is necessary for nations to continue to change.

It helps explain the transformation of societies from agrarian to early manufacturing, to chemicals and materials processing, and on to post-industrial or service economies through a variety of 'technology clusters' (Grubler 1994). Technological Innovation is also alleged to explain the different degrees of economic growth among industrialised countries through the 'Solow residual'. According to the standard interpretation, this residual may account for as much as half of the observed output growth and represents disembodied technological progress, usually referred to as total factor productivity.

Historically, advances in technology (1) were often concentrated in specific sectors, for example the use of fertilizers and pesticides in agriculture, or mass production in manufacturing, and (2) were sometimes deployed in many sectors, such as the harnessing of steam power, or the development of new materials such as plastics and ceramics. In the post-war years, there seemed no end to technological advancements, along with the jobs that they created. However in the 1970's, the overall rate of growth began to slow down and continued to slow down in the subsequent two decades. In the 90's, industries associated with the so-called knowledge-based economy began to grow and were responsible for an increasingly large share of employment growth.

It is argued that knowledge-based, information and communication technologies (ICT) have the potential to transform virtually every facet of production and consumption (OECD 1996). The microchip has doubled its information-processing capacity every 18 months (Mazurek 1998) and other dramatic changes occur with unprecedented speed. Beyond ICT technologies per se, it is argued that a ?knowledge-based? economy allows smarter production, products, and ways of working and doing -- and further, allows new ways of integrating heretofore segregated human activities. According to this view, knowledge-driven innovation will be the next engine of economic growth (Castells 1996; OECD 1996).

A somewhat contrarian view has recently been expressed by Drucker (1999). He argues that new technologies will indeed emerge, but they will have little to do with the "knowledge-based economy". He muses that e-commerce (electronic commerce), which will change the mental geography of commerce, will have the more profound effect by eliminating distance; there will be "only one economy and only one market."

Competition will know no boundaries, but the products and sectors that are affected will be eclectic and unexpected. "New distribution channels [will] change not only how customers behave, but also what they buy." And more to the point:

- The one thing...that is highly probable, if not nearly certain, is that the next twenty years will see the emergence of a number of new industries. At the same time, it is nearly certain that few of them will come out of information technology, the computer, data processing, or the Internet."

Drucker draws on both historical precedent for his predictions and on the observation that biotechnology and fish farming are already here. He opines that probably about a dozen technologies are now at the stage that biotechnology was 25 years ago. He reminds us that "the new industries that emerged after the railroad owed little technologically to the steam engine or to the Industrial Revolution in general," and that they were the product of a mindset that eagerly welcomed invention and innovation. Finally, he observes that

- "software is the reorganization of traditional work, based on centuries of experience, through the application of knowledge and especially of systematic, logical analysis. The key is not electronics; it is cognitive science. This means that the key to maintaining leadership in the economy and the technology that are about to emerge is likely to be the social position of knowledge professionals and social acceptance of their values."

Drucker also argues that this may require a radical change in the position of knowledge workers vis-à-vis their rewards and autonomy – and in industrial relations and labour policies.

Like ICT, biotechnology -- which, of course, is not a single technology -- has the potential for transforming agriculture, chemicals, pharmaceuticals, heath care, environmental cleanup, energy production, and even human reproduction itself (Krimsky 1982). New production methods and sources of food, chemicals, pharmaceuticals, and health care products are under development. Repair of undesirable genetic characteristics related to disease, the slowing of aging processes, the restoration of sight and hearing, and human reproduction are already the focus of research activity.

The transformation of unwanted byproducts of industrial production and waste, and creation of new sources of energy are already under development. Although many developments in biotechnologies have not advanced to the same extent as ICT, a revolution is in the making. In a way, biotechnology, too, should be seen as an important part of the expansion of a knowledge-based society -- since it is based on molecular biology rather than traditional chemistry -- although this is not what is traditionally meant by the term. In addition, initiatives such as the mapping of the human genome would not be possible without the prior revolution in information technologies.

New materials and 'nanotechnologies' are also already the focus of new research, and the interface between materials research and the life sciences has taken on renewed attention. These are some of the new technologies meeting the expectations of Drucker.

As industrial societies mature, the nature and patterns of innovation changes (Abernathy and Clark 1985, Utterback 1987). New technologies become old technologies. Many product lines (e.g., washing machines or lead batteries) become increasingly rigid, and innovation, if there is any, becomes more difficult and incremental rather than radical. In these product lines/ sectors, changes are focused on cost-reducing production methods -- including increasing the scale of production, displacing labour with technology, and exercising more control over workers -- rather than on significant changes in products. Gradually, process innovation also declines. A useful concept related to individual product lines is that of 'technological regimes', which are defined by certain boundaries for technological progress and by directions or trajectories in which progress is possible and worth doing (Nelson and Winter 1977).

Sometimes, the dominant technologies (such as the vacuum tube and mechanical calculator) are challenged and rather abruptly displaced by significant radical innovations (such as the transistor and electronic calculator), but this is relatively rare (Christensen 1997, Kemp 1994). As industrial economies mature, innovation in many sectors may become more and more difficult and incremental, regulatory and governmental policies are increasingly influenced, if not captured, by the purveyors of the dominant technology [regime] which becomes more resistant to change. However, occasionally, traditional sectors can revitalise themselves, such as in the case of cotton textiles.

Other sectors, notably those based on emerging technologies, may experience increased innovation. The overall economic health and employment potential of a nation as a whole is the sum of these diverging trends, and is increasingly a function of international trade. Whether nations seek to increase revenues based on competition in technological performance or alternatively on cost-driven strategies can have an enormous impact on employment (Charles and Lehner 1998).

As will be discussed below, health, safety, and environmental regulation, structured appropriately, as well as new societal demands can also stimulate significant technological changes that might not otherwise have occurred at the time (Ashford 1985). Within countries, we speak of 'national innovation systems' to denote the institutions, actors, and practices which influence innovation in general (Nelson and Rosenberg 1993, Nelson 1996). These include not only support for R&D; the training and education of scientists, engineers, entrepreneurs, and managers of technology; tax treatment of investment; regulations; and support of exports, but also networks involving trade associations, suppliers and other firms, customers, and workers. In the context of new growth theory, these networks are said to effectively compound accumulated knowledge and provide new energy for both growth and employment.

To the extent that innovation in the firm involves the increased participation of workers, and those responsible for innovation see workers as more than factors of production, worker know-how and creativity may yield unexpected benefits.

In addition to forces acting directly on the capacity to of a localised firm to innovate, an increasingly important additional driving force for growth is the globalisation of production and finance, which creates pressures, changes, and opportunities different from, and sometimes in contrary directions to, those created by technological innovation (Charles and Lehner 1998, Costanza and Daly 1991, Gordon 1995, OECD 1997).

In 1989, Mowery and Rosenberg argued that research and development, or more broadly, the accumulations of knowledge underlying technological innovation, appear increasingly to issue from global networks of enterprises and associated institutions.

This observation, echoing an increasing globalisation of commerce coincides with, but is not identical with the beginnings of the ICT explosion. Gordon, building on the optimism of Castells (1996), argues that globalisation enhanced by ICT "provides a basis for new forms of world-wide interaction and control, and liberates organizational structure from spatial constraints." His view of modern innovation is that it:

- Tends to be neither radical exogenous invention (as in the linear model) not narrowly path-dependent incremental change (as in the evolutionary model). Far more frequently, innovation tends to occur in the unilluminated space between these options: that is, while proceeding substantially within existing frameworks of knowledge and practice rather than initiating or requiring breakthroughs in science and technology, innovation nonetheless commonly tends to push at the margins of established organizational, technical, and economic practice as opposed cooperating within a more restricted field of "normal problem-solving routines." (Gordon 1995, pp. 180-181.)

Globalisation of production inevitably leads to a globalisation of the labour market, which in combination with a shift to a knowledge-based economy, raises concerns for wage fluctuations and employment (Freeman and Soete 1994, OECD 1996 and 1998).

EFFECTS ON ENVIRONMENTAL SUSTAINABILITY

Along with increases in the standards of living in developed countries, the unprecedented use of natural resources and energy, transformation of raw materials into products, and new agriculture, manufacturing, and production technologies are now known to both increasingly deplete the stock of resources and energy sources and to degrade the environment to the point that current industrial, agricultural, and transport systems are becoming unsustainable (Schmidt-Bleek 1998; Meadows et al. 1992).

We speak of undesirable ?negative externalties? in terms of the depletion of natural capital and compromises to environmental quality, harm to public health, and worker injury and disease.

The traditional ways of addressing pollution problems in terms of pollution control or so-called end-of-pipe approaches -- after technological systems are designed and implemented -- are no longer seen as adequate. Similarly, small advances in the efficiency of energy and resource use can no longer compensate for increased world demand and consumption of resource and energy-intensive technology. Even if we could somehow prevent traditional pollution and chemical accidents by the transformation to cleaner and inherently safer technologies, we would still be facing an increasingly unsustainable depletion of resources and the creation of greenhouse gases.

Energy, extraction, production, transportation and agricultural systems need to be inherently cleaner, safer, and resource conserving -- i.e., sustainable – in order to avoid or minimise depletion of resources and pollution. These systems need to be designed with the consideration of costs to the environment and consumer and worker health & safety in mind from the beginning, across every industrial sector, and in every function of the firm. Thus, rather than environmental technology, environmentally-sound and resource-conserving technology is needed, and this presents a challenge and opportunity for major technological innovation. But, as discussed earlier, more than incremental technological innovation is needed.

Radical and significant new approaches require that inputs and materials, final products, and processes be changed, but even more is needed. A shift to product-services with net significant dematerialisation is also needed, for example, through the leasing of carpets, washing machines, or automobiles with guaranteed maintenance or remanufacturing. Beyond product-services, entire systems may need to change, for example substituting transportation systems for individually operated automobiles or changing agricultural growing and distribution systems.

Technological development takes its cues from the market, and both societal/consumer demand and the regulatory environment shape the signals to which technology developers respond. This suggests that policies should be focused to influence consumers and technology providers to adopt sustainable practices. Dematerialisation, and shifts to product-services and beyond, requires technological, organisational, and social innovation to bring about the necessary physical and material changes, changes in infrastructure, and changes in social demands -- and to maintain flexible and learning institutions for continuous improvement.

EFFECTS ON EMPLOYMENT AND SOCIAL COHESION

A cleaner and less resource intensive environment is only one of several constituents of a sustainable society. Secure and meaningful employment, providing workers with adequate purchasing power, is an essential ingredient of a sustainable and socially cohesive economy. A growing economic system, one that increasingly satisfies human needs and wants (i.e., increases wealth), needs an adequate supply and quality of human capital. ICT and biotechnology are two technological newcomers that both challenge our conventional views of labour, production, and products -- and provide unanticipated opportunity for change. Whether these technologies as they are likely to develop will result in changes in the right direction remains to be seen.

The assertion that possible decreases in employment and/or wages brought about by labour-saving, productivity-enhancing technological change would be adequately compensated by lower prices, subsequent increased demand, and increased production volume is seriously being called into question (EC 1994, Freeman and Soete 1994, Head 1996, OECD 1996 and 1998). Incremental labour-saving innovation which dominates the majority of changes occurring in mature industrial economies is said to be at the root of creeping unemployment and underemployment involving the deskilling of at least some labour.

While new higher-skilled or newer-skilled challenging and rewarding work is being created in some firms or sectors, employment is being destroyed in others. It can not be said that the winners can compensate the losers in either the nature or the amount of employment.

Thus, in this scenario, net job creation is not an adequate metric of satisfaction with technological change. But further, the dichotomy itself may be far too simple. Gordon (1995) argues that both deskilling and reskilling can occur with similar technologies, task structures, and occupations, and that far from determining a unique outcome, information technologies simply expand the work organisation options.

The nature and rewards, both monetary and psycho-social, of work are undergoing structural change and revolution. But these changes are being brought about by new production, transportation, energy, and agricultural technologies that are undergoing innovation without concern or planning for their impact on the nature and level of employment. While compensatory policies, related to education, retraining, and the reorganisation of work exist or are being planned, they are reactive to technological changes. Here we need to take a lesson from the environmental problems created by rapid and extensive technological change. It is not sufficient to consider the effects on the environment as an afterthought. Environment quality needs to be built in. Similarly, it is suggested that thinking about work after technologies are planned and disseminated may be far too late to address their possible adverse consequences effectively.

We argue that production, consumption, environment, and employment ought to be co-optimized and considered simultaneously. This means technological, organisational, and social innovations need to be proactive and anticipatory, rather than reactive.

A knowledge-based economy potentially allows for more flexibility and new definitions of work, leisure, production, and consumption. The context established for innovations in all dimensions needs to reflect the realisation that the real wealth of the people lies in economic, environmental, and social sustainability.

Distinctions Between Sustaining and Disrupting Innovation

We have already discussed the need for a paradigm shift from the linear model of the innovation process (invention to innovation to diffusion) – to an iterative, network-influenced model to explain the back-and-forth of scientific and engineering advances -- and the influences on innovation upstream and downstream in the supply chain.

A further paradigm shift is needed to explain why product-oriented firms that listen closely to their customers can in some cases succeed impressively, and in other cases fail when a new entrant introduces a product that literally destroys their market. This is important to understand if we are to have any hopes of influencing private sector activities in the direction of sustainability.

For this purpose, a distinction needs to be made between the description of product innovations as incremental or radical, and product innovations as sustaining versus disrupting. Sustaining innovations occur by established firms pushing the envelope to continue to satisfy existing consumers with improved products. Those improvements may be incremental or radical, and come in successive waves by established firms in that product market. Disrupting innovations cater to different, perhaps not yet well-defined, customers with product attributes different from those in the established producer-consumer networks. Ultimately, they may displace the established technology and market.

Christensen's (1997) concept of a 'value network" is "the context within a firm identifies and responds to customers' needs, solves problems, procures input, reacts to competitors, and strives for profit". In principle, product attributes related to triple sustainability could be important to some customers, for example, resource intensiveness or the way the products are made. An emerging case in point is environmentally friendly packaging that appeal to a defined customer base.

Product attributes are valued differently by different value networks. Existing mainstream customers may demand different things than 'special customers' who are presently small in number, but who could eventually reflect future mainstream demand. Sometimes networks emerge that reject a product previously accepted. Producers of genetically-engineered foods were reinforced by their traditional consumer network that these foods would be acceptable, and they therefore ignored a small, but vocal and different group of consumers who ultimately became a serious force to contend with. The industry was lulled into complacency because they surveyed and listened to their main customers and did not entertain the possibility that things would change.

Disrupting innovations are technologically straightforward, often consisting of off-the shelf components put together in a [new] product architecture that is often simpler than prior approaches. They could, but need not be, heavily R&D driven. They offer less of what customers in established markets want and hence can rarely be employed there.

The firm's organizational structure and the way its groups learn to work together can then affect the way the firm can -- and cannot -- design new products. Managerial decisions that make sense for companies outside a value network make little sense for those within it, and vice versa.

In addition to shifting consumer demands, regulation can 'make a market' by providing directions for technological change and product attributes. Of course, regulatory requirements that are viewed as disruptive by the firm and its managers often require disrupting technological changes -- and that is why managers of established firms that pursue sustaining innovation resist regulation and will try to influence regulation that can be satisfied by sustaining innovations -- if not by diffusion of their existing technologies.

Christensen (1997) discusses what could characterize the [rare] successful management of disruptive innovation by the dominant technology firms, rather than give way to their displacement by new entrant firms.

- Managers align the disruptive innovation with the 'right' customers.
- The development of those disrupting technologies are placed in an organizational context that is small enough to get excited about small opportunities and small wins, e.g., through 'spin-offs' or 'spin-outs'.
- Managers plan to fail early, inexpensively, and perhaps often, in the search for the market for a disruptive technology.
- Managers find new markets that value the [new] attributes of the disrupting technologies.

Since, this is rarely done in the commercial context of product competition, it is unlikely to occur for many sustainability goals without either strong social demand or as a result of regulation. This reinforces our view that disrupting innovations are necessary and the policy instruments chosen to promote triple sustainability need to reflect these expectations.

Requisites for Technological and Organisational Innovation

In industrial economies, the firm is the most important locus of technological innovation, although, as mentioned above, the construct of ?innovation networks? involving suppliers, consumers, workers, trade associations, others firms, and government more accurately captures the dynamics of the innovation process. In addition, government itself historically has also had an important role to play as a direct source of innovation, especially in the area of ?big science? such as in the cases of the early development of computers, air transport, and cancer therapies. The term 'innovation systems' has be described as "a set of institutions whose interactions determine the innovative performance...of national firms" (Nelson and Rosenberg 1993).

Sometimes science and understanding precedes application (engineering), as in the case of modern chemical technologies. Sometimes applications precede and understanding follows, as in the case of electrical equipment industries. As discussed earlier, a better description of the innovation process might be an iterative process where understanding and application provide dynamic feedback to one another, giving rise to stages of advancement.

In order for innovation to occur, the firm (or government itself) must have the willingness, opportunity, and capability or capacity to innovate (Ashford, 1994). These three factors affect each other, of course, but each is determined by more fundamental factors.

Willingness is determined by both (1) attitudes towards changes in production in general and by (2) knowledge about what changes are possible. Improving the latter involves aspects of capacity building, while changing the former may be more idiosyncratic to a particular manager or alternatively a function of organizational structures and reward systems. The syndrome ?not in my term of office? describes the lack of enthusiasm of a particular manager (in the firm or government agency) to make changes whose benefit may accrue long after he has retired or moved on, and which may require expenditures in the short or near term.

In the context of disrupting innovation by firms representing the dominant technology, willingness is also shaped by the [rare] commitment of management to nurture new approaches that are at odds with its traditional value network. In instances where firms pursue a dual strategy, the new initiatives are often short-changed in terms of the allocation of resources and top-flight researchers (Christensen, 1997).

Opportunity involves both supply-side and demand-side factors. On the supply side, technological gaps can exist (1) between the technology used in a particular firm and the already-available technology that could be adopted or adapted (known as diffusion or incremental innovation, respectively), and (2) the technology used in a particular firm and technology that could be developed (i.e., major or radical innovation). On the demand side, four factors could push firms towards technological change -- whether diffusion, incremental innovation, or major innovation -- (1) regulatory requirements, (2) possible cost savings or expansion of profits, (3) public demand for more environmentally-sound, eco-efficient, and safer industry, and (4) worker demands and pressures arising from industrial relations concerns. These latter two factors could bring about changes in the value networks, and could stimulate change too late in the dominant technology firms, if new entrants have already seized the opportunity to engage in developing disrupting innovations.

Capacity or capability can be enhanced by (1) increases in knowledge or information about more sustainable opportunities, partly through deliberately undertaken Technology Options/Opportunity Analyses, and partly through deliberate or serendipitous transfer of knowledge from suppliers, customers, trade associations, unions, workers, and other firms, as well from available literature, (2) improving the skill base of the firm through educating and training its operators, workers, and managers, on both a formal and informal basis, and (3) by deliberate creation of networks and strategic alliances not necessarily confined to a geographical area or nation or technological regime. Capacity to change may also be influenced by the inherent innovativeness (or lack thereof) of the firm as determined by the maturity and technological rigidity of particular product or production lines (Ashford et al. 1985; Ashford 1994).

The heavy, basic industries, which are also sometimes the most polluting, unsafe, and resource intensive industries, change with great difficulty, especially when it comes to core processes. However, new industries, such as computer manufacturing, can also be polluting, unsafe (for workers), and resource and energy intensive, although conceivably they may find it easier to meet environmental demands. It deserves re-emphasizing that it is not only hardware, materials, process, and product technologies that are rigid and resistant to change.

Personal and organisational flexibility are also important. Technology, organization, and people combine together to influence change. The willingness, opportunity, and capability/capacity to change all three must be addressed in order to encourage sustainable development.

Finally, it is important to realise that Factor 10 (or greater) improvements require radical shifts in production, use, and consumption patterns and are not tied to any particular technological fix. Shifts to product-services which shift the focus from the production of products to the delivery of functions and utility (services) are central to dematerialisation, energy de-intensification, and the creation of new employment. Programmes and Instruments for Stimulating Technological and Organisational Innovation

Having addressed the importance of willingness, opportunity, and capability/capacity to undergo technological and organisational change, this section discusses the possible role of support for national innovation systems, regulatory intervention, and other instruments for stimulating and directing innovation through their influences on willingness, opportunity, and capability/capacity to innovate.

NATIONAL INNOVATION SYSTEMS

Nelson and Rosenberg (1993) define an innovation system as "a set of institutions whose interactions determine the innovative performance...of national firms." All industrial countries use government policies and programmes to stimulate technological innovation (Allen et al. 1978, Nelson and Rosenberg 1993, Nelson 1996). These government initiatives have historically focused either on enhancing infrastructure and the viability of the economic system, or in some cases have been focused on specific sectors. For example, in the case of addressing disease, initiatives have involved support for the development of pharmaceuticals, especially for widespread difficult diseases such as cancer and for so-called 'orphan diseases' where the potential market is small because a small number of people are affected. While today government policies do not directly intervene in the development of ICT technologies, it should be remembered these technologies benefited from targeted developments associated with national defense concerns in the United States.

The standard activities undertaken by nation states in furtherance of industrial development consist of direct (or indirect) support of R&D in industry, universities, and through government research institutions and laboratories -- and the training of scientists and engineers.

Inasmuch as successful innovation requires much more than R&D, national innovation systems also include support for institutions that educate, train, and retrain workers; provide a system of industrial relations and dispute resolution; finance investment and development; oversee the regulation of health, safety, and environment, and provide the general infrastructure for an industrial economy to function effectively in both national and international markets.

From out earlier discussion of the nature of the innovation process, it is worth remembering that science (discovery) and engineering (application) are intertwined and that a linear model of R&D leading straightforwardly to development and application does not capture the iterative, back and forth dynamics of innovation.

In many cases new scientific understanding follows, rather than leads application. Furthermore, the classic Schumpeterian innovator – the first to bring a new product to the market -- is frequently not the firm that benefits mostly from the innovation (Nelson and Rosenberg 1993). The role of outsiders in a traditional technological regime can be particularly important (van de Poel 1999). Thus, institutional factors which promote feedback, learning, and dynamic interaction among scientists, inventors, innovators, engineers, suppliers, customers, regulators, and other actors are all a crucial part of the innovation system.

Some countries, like Japan, have specific targeted industrial policies. Contradicting the conventional wisdom that liberalised markets are the most important determinants of development, Amsden (1994) argues that it is precisely targeted governmental policies that underlie the successes of the 'Asian Miracle'. Yet, many economies, like the United States, focus more on creating an economic and regulatory environment conducive for innovation. Of special importance is the realisation that as markets and firms have become more globalised, the appropriateness of an innovation system to be described as 'national' has been questioned (Nelson and Rosenberg 1993) and along with this realisation that national policies can continue to matter as much as they did historically.

Studies of various national systems yield mixed results for the effectiveness of government policies (Amsden 1994; Ashford 1976; Freeman 1988; Hill and Utterback 1979; Nelson and Rosenberg 1993). For so-called 'big science projects', like the early development of computers and aircraft, government did indeed make a positive difference. For more usual innovation carried out by industrial firms, the story was mixed. A study of innovation in Europe and Japan in the late 1970's (Allen et al. 1978; Ashford 1976) conducted by the Massachusetts Institute of Technology (MIT) found that the only significant (and positive) effect of government policies came from health, safety, and environmental regulation -- a result that was counter-intuitive at the time.

The MIT researchers did not attribute the absence of findings of government effectiveness so much to the failure of governments to pick winners, but rather to perverse incentives faced by national government bureaucrats managing new technology programs to prove success to their superiors by supporting 'sure-thing' projects that turned out to be of little technological importance (Ashford 1976).

A recent survey of 716 companies in 17 countries in Europe regarding government policies for innovation, including R&D grants, educational programmes, or tax breaks, found that success in product innovation could not be linked to the overall environment for innovation in the country (Marsh 1999).

The Financial Times reports that "[T]he study suggests that while broadly based government measures to alter the economic climate for innovation may not work, there is a place for more targeted schemes to encourage companies at a 'micro' level in areas linked to the innovative process" (Marsh 1999). The concern in this work is not to settle whether or not support for national innovation systems can stimulate innovation as a general matter, but rather to enquire into possible governmental roles in stimulating technological innovation (and diffusion) to achieve triple sustainability. Relevant to this question is what effects government intervention might have on innovation that affects environment and employment.

The Effects of Environmental, Safety, and Health Regulation on Technological and Organisational Innovation

The early MIT research mentioned above (Ashford 1976) stimulated more focused research into the effects of government regulation in the United States (Ashford et al. 1984) which found in a number of MIT studies beginning in 1979 that regulation could stimulate significant fundamental changes in product and process technology which benefited the industrial innovator, as well as improving health, safety, and environment, provided the regulations were stringent, focused, and properly structured.

This empirical work was conducted fifteen years earlier than the emergence of the so-called Porter Hypothesis which argued that firms on the cutting edge of developing and implementing technology to reduce pollution would benefit economically by being first-movers to comply with regulation (Porter 1990, Porter and van den Linden 1995).

One could describe the Porter Hypothesis as having a weak and a strong form. Porter himself actually discusses only the weak form. The weak form is essentially that regulation, properly designed, can cause the [regulated] firm to undertake innovations that not only reduce pollution -- which is a hallmark of production inefficiency -- but also save on materials, water, and energy costs, conferring what Porter calls 'innovation offsets' to the innovating firm. This can occur because the firm, at any point in time, is suboptimal. If the firm is first to move by complying in a clever way, other firms will later have to rush to comply -- and do so in a less thoughtful and more expensive way. Thus, there are "learning curve" advantages to being first and early. Porter argues that in the international context, first-mover firms benefit by being subjected to a national regulatory system slightly ahead of that found in other countries.

What is missing from Porter's analysis are details about the process of innovation, how change actually occurs in industrial firms, what kinds of firms are likely to come up with what kinds of technical responses, and how very stringent regulation can confer competitive advantage beyond what he calls 'innovation offsets'.

While Porter stresses the importance of going beyond the 'static model' of compliance responses, he is in fact talking about modest or incremental innovation in pollution control, and to a lesser extent significant pollution prevention.

The strong form of the Porter Hypothesis was not put forth by Porter at all. It (and the weak form as well) was first proposed by Ashford and his colleagues at MIT after years of cross-country and US-based studies that showed that stringent regulation could cause dramatic changes in technology, often by new firms or entrants displacing the dominant technologies. The replacement of dominant technologies by new entrants, rather than incremental change by existing technology providers, has been the source of the most important radical innovations over this century.

It stands to reason that any strong change in market conditions -- be it sudden factor cost changes, new opportunities from new consumer or societal demands, an energy crisis, or demanding regulation -- could stimulate significant innovation.

With regard to regulation, what seems to matter is not only the stringency, mode (specification versus performance), timing, uncertainty, focus (inputs versus product versus process) of the regulation, and the existence of complementary economic incentives -- but also the inherent innovativeness or lack of it by new entrants or regulated firms (Ashford and Heaton 1983, Ashford at al. 1985).

There is a rich literature providing examples of more and less stringent regulation with predictably more or less innovative responses (Strasser 1997). A consistent behavioral theory emerged after nearly twenty years of work at MIT and it is useful for policy design (Ashford et al. 1985, Ashford 1993,1994, and 1999). This theory requires an understanding of both the importance of getting the various components of the regulatory signal just right, and the different innovative potential of both existing firms and new entrants for product and process innovation. The importance of new entrants is missing in the analysis offered by Porter.

It is useful to trace the intellectual thinking on the regulation-innovation interaction. MIT research, which began in the 1970s as a cross-country investigation of the effects of government policies on innovation in France, Germany, Holland, the UK, and Japan, found paradoxically that the only government policy that affected innovation was in fact health, safety and environmental regulation, not strategies government devised as a part of its industrial policy (Allen et al. 1978). Moreover, the effects of regulation on innovation turned out to be positive, not negative as expected by the conventional wisdom at that time. This unexpected observation prompted a more detailed examination of regulation in the chemical, pharmaceutical, and automobile industries from which the MIT behavioral model emerged (Ashford et al. 1979, Ashford and Heaton 1983).

In 1979, the MIT researchers argued that "ancillary benefits" (what Porter called "innovation offsets" more than a decade later) could yield costs savings to the regulated firm. MIT researchers argued that "a study of the innovation process in five foreign countries found that innovations for ordinary business purposes (not necessarily for compliance) were much more likely to be commercially successful when environmental, health, and safety regulations were present as an element in the planning process than when they were absent". They further argued that "regulation, by adding new dimensions to older problems, increases the problem space of the engineer". Distinctions were made between innovative technology for compliance purposes and main business innovation. Regulation could confer ancillary benefits, but more significantly it could rechannel creativity, bring new skills, and cause beneficial reorganization in the firm. This was the precursor to the soft version of the Porter Hypothesis.

Later in 1983, as a result of intensive study in the chemical producing and using industries, MIT argued that stringent regulation could stimulate entirely new products and processes into the market by new entrants with the displacement of dominant technologies rather than the transformation of technologies by existing firms. One of several vivid examples is the displacement of Monsanto's PCBs in transformers and capacitors by an entirely different dielectric material pioneered by Dow Silicone.

In regulatory systems where health or safety concerns are sufficiently serious to counteract industry pressure, there is plenty of technology innovation forcing (OTA 1995). Often this regulation involves dramatic reductions to workplace exposure, consumer product bans, or the significant reduction of industrial emissions or effluents or the banning of industrial products. The point is that regulation can be designed to stimulate radical innovation if there is both sufficient social concern and political will. This is especially true where regulatory goals are clear and demanding, but the means of complying are flexible. Nor is true that negotiation, rather than regulation yields superior technology and more protection (Gouldson and Murphy 1998, Caldart and Ashford 1999).

Supplementing the earlier US evidence that regulation can stimulate technological changes that both benefit the innovators and improve safety, health, and environment is a recent report from the Stockholm Institute comparing pre- and post regulatory environmental compliance costs of industry, and documenting the stimulation of technology by regulation in both Europe and the United States.

The report complements other empirical data in the United States showing regulatory stimulation of cheaper and innovative technological responses both in the environmental area (Strasser 1997) and in the area of worker health and safety (OTA 1995).

Regulation can thus encourage disrupting innovations by giving more influence to new 'value networks' in which demands for improvements in both environmental quality and social cohesion are more sharply defined and articulated. Of course, industries that would fear disrupting new entrants would not be expected to welcome this regulation. This explains in part their resistance to regulation and their propensity to try to capture regulatory regimes, surreptitiously or through direct negotiation (Caldart and Ashford 1999).

TECHNOLOGY AND EMPLOYMENT

The OECD has been particularly active in researching the connections between policies that focus on innovations in the post-industrial knowledge-based economy and employment (OECD 1996 and 1998), again emphasising the importance of networks which can efficiently distribute knowledge and information.

It argues that the knowledge-intensive or high-technology parts of the economy tend to be the most dynamic in terms of output and growth. Not surprisingly, the OECD (1996) identifies the priorities to be: (1) enhancing knowledge diffusion, especially by broadening support of innovation from "mission-oriented" science and technology project to "[knowledge and technology] diffusion-oriented" programmes, (2) upgrading human capital, and (3) promoting organisational change. In its 1998 job strategies report, OECD (1998) goes into considerable details on specific policies. Most are focused on liberalising markets, forging stronger or new partnerships, and involving key stakeholders. None directly address anthropocentric production or the design of meaningful, remunerative employment. Employment is derivative of the creation of industrial technology.

It is hard to argue with most of these recommendations, with the exception of liberalising labour markets. Allowing wages to fall is a solution rejected by the European Union in its white paper on Growth, Competitiveness, and Employment (EC 1994). Kleinknecht (1998) offers a powerful argument in defense of not reducing wages to accommodate unemployment. Drawing on evidence from European labour markets, he maintains that while advantageous in the short run, from a Schumpeterian perspective, allowing wages to fall to accommodate the less innovative firms who do not use labour or capital effectively will result in less product and process innovation in the long run.

The author argues that the process of 'creative destruction' advocated by Schumpeter should be allowed to occur, lest we encourage suboptimality. Essentially, this argues for forgoing immediate gains from achieving static efficiency in order to realise a better dynamically-efficient outcome. Charles and Lehner (1995) similarly argue that labour market liberalisation can remove the pressure on firms to pursue productivity growth [most importantly, by using labour more effectively].

Kleinknecht argues that although there has been a slower growth in labour productivity in the Netherlands due to a slower adoption of new process technology, there is likely to be a greater focus on product innovation which is expected to create more jobs and jobs of better quality if wage flexibility measures are not adopted. Furthermore, the process of innovation will stimulate more firm-specific [tacit] knowledge which will have a cumulative, multiplier effect on future innovativeness. In Kleinknecht's view, the longer-term contribution to the economy will more than offset the disadvantages of a longer [old] product life brought about by keeping capital stock around for a longer time.

Furthermore, if allowed to operate, the winds of creative destruction will remove the less innovative firms from the market. In contrast, lowering wages for less innovative firms will enable those firms to remain and benefit from their short-term economic advantage.

In discussing the recent economic situation in the Netherlands, Kleinknecht sees problems ahead because of recent lower wage increases that will manifest in contributing to a lowering of growth in labour productivity, in the quality of entrepreneurship, and in lower purchasing power and [consumer] demand. Also problematic are attempts to lower labour protection policies. Greater latitude in firing workers would give a competitive advantage primarily to non-innovators and at the same time lower the commitment of workers to the enterprise.

INSTRUMENTS TO ENCOURAGE TECHNOLOGICAL INNOVATION FOR TRIPLE SUSTAINABILITY

The policy maker faces a dilemma. Since the evidence for success of targeted governmental policies to direct technological change is mixed – and in the face of evidence that technological innovation is a complex and changing phenomenon, especially, in the context of both globalised trade and the knowledge-based society, it might be suggested that the proper role for government vis-à-vis innovation is to establish a conducive environment, but to avoid targeted policies.

Against this line of reasoning is the realisation that technological change has brought and continues to bring adverse effects to the environment in terms of resource depletion, energy use, and pollution -- and adverse effects to employment in terms of level, nature, skill content, rewards, and dislocation. Further there is a kind of cognitive dissonance between policies that seek to move the economic from suboptimality to static efficiency – and those that seek new dynamic equilibria, including incentives for continuous improvement.

This suggests that government ought to have a strong presence, not only to "internalise the social costs of production and commerce (where there are market failures), but also to transcend markets (where even perfectly working markets are inadequate). Although, the ideology of laissez faire suggests that government regulation is mostly unhelpful or is inefficient, there is increasingly persuasive evidence that regulation – properly designed – is not only necessary to achieve sustainable economies – but that it can actually stimulate innovation leading to improved competitiveness, an improved environment, and employment.

The challenge, then, is how to engineer a coherent and comprehensive strategy. Echoing a previous observation that there are many more ways to get it wrong than to get it right, one place to start is to remove the perverse incentives in the system. These perverse incentives include the failure to account for resource depletion in the cost of materials and energy, taxing the "goods" like labour instead of the "bads" like pollution, subsidising the continued existence of old firms and technologies, rewarding firms that underutilise human capital by allowing wage flexibility and easing labour protection laws, and a host of other policies.

A targeted approach using both carrots and sticks requires a better understanding of what is required for firms to innovate – that is, the willingness, the opportunity, and the capacity/capability to innovate.

In addressing strategic design options for innovation that addresses a particular problem area or exploit an curative opportunity, we need to ask four key questions:

- What are the characteristics of needed innovation that are desirable (environmental aspects; resource and energy intensiveness; whether process, product, process, service aspects or whole systems need to be changed; associated worker hazards; embodiment of knowledge; associated job skills; employment aspects, etc.)?
- For whose benefit is the innovation (manufacturers, service providers, industrial users, consumers, workers, others)?
- Who/what sector is likely to innovate (existing firms, new entrants, government, universities, research institutes, etc.)?
- How can this innovation be brought about?

Instruments for stimulating technological and organisational innovation affect one or more of the necessary characteristics -- willingness, opportunity, and capability -- and include those that influence nationally-focused supply and demand-side factors and more general features of a nation?s infrastructure.

In the context of increasingly globalised economies, instruments which establish a level playing field and minimize the 'tragedy of the commons' and the free rider problem are needed (Costanza and Daly 1991, Krugman 1986). These include multilateral environmental agreements, labour protection conventions (such as those of the International Labour Organisation), and the rules of international trading regimes such as the World Trade Organisation (WTO) implementing the General Agreement on Tariffs and Trade (Folsom et al. 1996).

It must be realised that there are four different kinds of barriers to the emergence of sustainable economies. The first is traditional market imperfections, which include mostly unintended environmental and human consequences (externalities) of the industrial production and consumption system.

The second arises because even perfectly working markets are inherently imperfect where (1) distributional impacts within and between nations are of central concern, (3) the value of achieving sustainability benefits occurs far into the future and does not have a sufficiently large present economic value, disadvantaging the avoidance of future harm and/or generations, and (3) achieving dynamic efficiency requires sacrifice of static efficiency in the near-term. Getting the prices right will not overcome these problems. Longer time horizons needed for action.

A third barrier is the absence of a well-defined, visible, and influential [value] network that represents the preferences for a sustainable society. A fourth barrier is that political leadership and institutions do not necessarily maximize net welfare even in the short-term, but instead cater to special interests not particularly interested in promoting triple sustainability. Paradoxically, even where there are economically efficient markets, they can ignore distribution of wealth and purchasing power, which can result in socially unsustainable societies.

We argue that government can have a major role in making a market for the appropriate kind and level of innovation (Ashford 1985), for example by correcting market imperfections, and can transcend markets altogether in other instances, for example where distributional or equity concerns within and between generations or between nations are of primary interest and markets are incapable of addressing these concerns.

Instruments applied at the national level which can help circumvent these barriers to long term sustainability include those which focus on:

- Getting the prices right by appropriate financial treatment of natural capital (Costanza and Daly 1991), by internalising social costs in the costs of pollution (Baumol and Oates 1988), for example through pollution taxes, and by removing taxes on labour.
- Reforming the health, safety, environmental, and energy regulatory system so that regulations can stimulate sustainable innovative technological and organisational changes through the creation of stringent, clear, and certain targets (Ashford 1994 and 1999).

This is preferable to freezing technological advance resulting from regulatory capture and timid government leadership through lax, unclear, or uncertain standards and market signals encouraging the diffusion of inadequate existing technologies and approaches.

- Changing the tax treatment of investment [for example, by preferentially promoting technologies/approaches which reduce resource/energy use (thereby increasing resource/energy productivity)]

- Increasing the rewards to firms for creating meaningful and remunerative work, through both favourable tax treatment of investment for the creation and adoption of job creation and job content-enhancing technology, as well as through decreasing the taxes on labour.
- Preventing wages from falling to accommodate unemployment.
- Implementing programmes for the education and training of labour for a knowledge-based economy.
- Implementing programmes for the education of technical and professional designers and managers for the knowledge-based economy (the universities have a particularly important to play here), and
- Building partnerships with proactive, technology-leading companies, but taking care to avoid "capture".
- Encouraging or requiring management to bargain with workers before technological changes are planned and implemented (Ashford and Ayers 1987).

These measures include both carrots and sticks, and include both regulatory and economic instruments, and long-term as well as shorter-term instruments. (For an excellent discussion of detailed progammatic elements and policy instruments, see OECD 1994 and 1995).

They are some of the measures that comprise what we would call an industrial policy for triple sustainability – one that guides both the technology providers on the supply side and influences the kinds of processes, products, services and work demanded by consumers, society, and workers.

This policy has to be both short and longer range and has to represent the interests of both existing firms, workers, consumers, and citizens – and those of future stakeholders. Value networks that give rise to disrupting innovations do represent near-term future interests, but more is needed, especially in mature industrial systems. The government should act as a trustee for the health of its future citizens and workers, the future environment, and the technology of the future.

As helpful and as necessary as the existing or emerging firms are, they can not be the sole trustee of future interests. Only government can do that. As already discussed, in some cases, government needs to make a market for innovation; in others, government must transcend markets altogether.

Instruments for Stimulating Social Innovation

More than technological and organisational innovations are needed. Innovations in social attitudes, communication networks, and lifestyle which affect both demand and supply are also crucial. If there are limits to growth, then there are also practical limits to social choices. We are not arguing that government should limit social choices per se, but rather that government should expand choices into more sustainable options through technological, organisational, and social innovations to encourage socially responsible, informed choices. The instruments available for encouraging social innovation are educational, economic, and legal or regulatory in nature.

Already mentioned are ?innovation systems/networks? which can change what and how industrial users and consumers express their demands and concerns to product and service providers. It was argued earlier that con¬sumers and producers must come together at a very early stage of the innovation process (possibly through the creation of innovation teams). Innovations, in order to be successful, as a rule increasingly start with organizational and social innovations.

Education and the provision of information about the need for sustainable production and service systems, and about the options that could and need to be developed, can influence industrial use, consumer, and worker demand for the satisfaction of basic needs, wants, and the use of leisure or saved time.

Public participation in both private-sector decisions (Irwin et al. 1995), usually though networks involving technology providers and suppliers, and in governmental decisions (Sclove 1995), through informal networks or more formal involvement through advisory panels and science shops, may have more direct and influential effects than through the purchasing power of consumers – especially where the problems relate to material, product, and energy transfers among industrial sectors (See the earlier discussion on value networks).

In addition, unlike the relationship between producers and consumers which is subject to change as consumers change allegiances and preferences for a particular product or service provider, industrialists and labour have an intricate relationship involving an explicit or implied employment contract, job health and safety, other worker safeguard legislation, and frequent, if not daily contact.

Their relationship is influenced by both a complex web of laws and by industrial custom (Ashford and Caldart 1996). Through "technology bargaining" between management and labour (discussed more fully in the previous section) workers could make some of their needs and wants known to management and possibly influence innovation that affects working conditions, the products or services that the firm offers to consumers and industrial customers, and the resource and environmental consequences of its activities.

In practice, management usually holds quite tenuously to its prerogatives to make unilateral decisions concerning changes in the technology or technical trajectory of the firm. Labour and industrial relations law protects this management prerogative to various degrees, depending on the country. To the extent that decisions affecting technology are shared, what the firm produces and how the firm functions could very well change. This social innovation would require both legal/institutional and cultural changes. The co-dependence of workers and owners of industrial enterprises for evolving into firms that are both competitive and sustainable could become even more important with increased globalisation of industrial economies.

Finally, changing the nature and rewards of employment through a responsive industrial relations system could, in turn, indirectly affect the level and character of consumption that workers desire or need in their capacity as consumers.

Trade as a Driving Force Affecting Sustainability

While technological innovation has been described as the traditional engine of economic growth, trade and globalisation are increasingly being pursued by firms and nations as a means to increase revenues, offering global consumers lower prices as a result of lowering factor costs and taking advantage of economies of scale.

Neoclassical trade theory (Krugman 1986) extends to the international arena the theory of comparative advantage -- whereby mutually advantageous bargains could be struck in self-contained economic systems (Riccardo 1817). This, and the influence of competition among producers to provide a variety of goods and services at reasonable prices for the society, are the bases for the enthusiasm for market economies, for commerce both within and among nation states.

Of course, when production occurs by creating human and environmental externalities -- such as damage to health and ecosystems -- regulation or some other mechanism such as taxes is seen as a way to internalise the uninternalised social costs. However, not all nations do this, or do this to the same extent. All developed countries have instituted programs to address these problems. However, with the advent of an expanded globalisation of trade in inputs, resources, materials, products and services – especially involving developing countries-- new problems have arisen.

These problems relate to both differences in what different nations trade and in the increasing tendencies to externalise the social costs of production by offering little effective environmental and labour protection in order to keep costs low.

Trade between nations with different material, resource, skill, and cultural endowments at very different levels of economic development, of course, encourage trading on the bases of the comparative advantages of the trading partners (Krugman 1986). However, the value of, and hence price and revenues associated with, what each country produces or sells depends on many factors. Recent history reveals that the developing countries which sell basic materials, food, and other commodities experience continuing falling demand and hence revenues, while developing countries selling primarily manufactured goods and knowhow experience continually increasing demand and revenues for their traded goods. This has lead to an increasingly widening gap in the trade surpluses/deficits between developing and developed countries (Daly 1991 and 1993).

The problem had become so serious that third world countries can not even finance their external debt. In contrast, the so-called Asian tigers learned early to invest in knowhow and technology -- with the help of deliberate government industrial policies -- and became heavy competitors with the developed countries in finished products and manufactured goods. The changing economic positions of both developed and developing countries have different, but adverse consequences for environment and employment. In the wealthy countries, more consumption is encouraged and unless the nature of consumption changes, this means an increased load or rucksack on the environment.

In the poorer countries, financial resources become increasingly unavailable to address environmental problems and to create employment with adequate purchasing power.

But these problems have an additional wrinkle. To the extent that nations trade on uninternalised externalities (i.e., they do not internalise the human, social, and environmental costs of production), the amount of resources, energy, pollution, and waste produced is both economically and ecologically inefficient. This means that too much is produced, used, and disposed of. To the extent that producers increasingly compete on the basis of cost -- or cost reduction -- the more the social and human costs are likely to be externalised.

Trade based on performance advantages, for which continual innovation is required, can be replaced with trade based on cost, and cost reduction, where in addition to becoming less sustainable, trade relies on using existing plant and technologies rather than on technological innovation. Thus, in some cases, trade can be the enemy of innovation -- and it is innovation that is required to achieve sustainabilty.

To the extent that multilateral environmental agreements and international labour conventions are signed by trading partners so that similar environmental and labour standards are adopted, implemented and enforced, trade advantages will not hinge on differences in these standards (Ashford 1997). However, many international accords, though signed, have not been ratified or implemented. And if implemented, they have often not been enforced, nor is compliance widespread.

Some advocates of free trade (Bhagwati, 1993 and 1997) argue that nations ought to be encouraged to trade on these differences in environmental protection and labour standards. Where countries have not signed, ratified, or implemented environmental, product safety, or labour standards, the rules of trade are governed mainly by the trade regime in which they participate, such as the General Agreement on Tariffs and Trade (GATT), the North American Free Trade Agreement (NAFTA), or the Association of Southeast Asian Nations (ASEAN).

Unfortunately, In the context of the GATT, in the absence of a situation where the trading partners are otherwise bound to international environmental and labour accords, the only exception allowing the prohibition of trade (through import bans or taxes) for environmental and health reasons in Article 20 of the GATT is that the importation will hurt the health or environment of the importing nation (Ashford 1997, Folsom et al. 1996).

Under the GATT, if the exporting country spoils its own environment or exploits its own workers? health and safety, the importing country can not retaliate. So-called PPM?s – 'process and production methods' – are considered the exporter's own business. While it is plain to see that production which is cheaper because an exporting firm does not care for the environment or worker health and safety is in fact a subsidy to production on the back of its nation's environment and workers, this is not the kind of subsidy -- like is financial support from the government -- that the GATT prohibits.

Obviously, there is a need for strengthening participation in, and for coordinating, multilateral environmental, labour, and financial capital agreements (Ehrenberg 1996). In addition, there needs to be an appropriate incorporation of environment, labour and social concerns in world trading regimes. But this remains a difficult goal to achieve. There is resistance to 'greening the GATT' from both developed and developing countries. While both Europe and the United States want to include labour issues in trading regimes, developing countries and Japan are adamant about keeping labour issues out. The side agreements of NAFTA on environment and labour are an exception in trading regimes (Folsom et al. 1996).

In the meanwhile, at the national level economic, environmental, employment, and trade policies also need to be coordinated. Fortunately, within some trading blocs like the European Union, uniformity is a more of goal and minimum standards are almost always higher than the lowest standards of its members.

Globalised trade outside the European Union creates special pressures on competition and employment too, and increased national revenues of one trading partner usually arise from either superior quality of goods and services, or from comparative advantages in factor prices through cheaper labour and exploitation of the environment and resources. To the extent that all nations do not equally internalise the social costs of production, cost advantages can dominate over product performance.

Thus, in some circumstances, trade can compromise the incentives for environmentally-sound, job-enhancing technological change. Thus, the market and social opportunities for technological innovation need to be clearly defined and understood. Here, too, incremental innovation may not suffice. A bolder vision is necessary. Finally, the distribution, as well as levels of wealth -- broadly defined -- needs particular attention in the shaping of policies. This requires strategic choices that reflect a concern for intergenerational equity, equity within nations, and equity among nations (Sen 1992).

While this volume focuses on changes that can be made first within the industrialised nations, it must be realised that both technological innovation and trade will affect both industrialised and developing nations, although in different ways. Thus, economic, environmental, employment, social and trade policies should work in tandem to move all societies in a more sustainable direction.

SUSTAINABLE DESIGN

Sustainable design (also called environmental design, environmentally sustainable design, environmentally conscious design, etc.) is the philosophy of designing physical objects, the built environment, and services to comply with the principles of social,economic, and ecological sustainability.[1]

THEORY

The intention of sustainable design is to "eliminate negative environmental impact completely through skillful, sensitive design".[1] Manifestations of sustainable design require no non-renewable resources, impact the environment minimally, and connect people with the natural environment.

Beyond the "elimination of negative environmental impact", sustainable design must create projects that are meaningful innovations that can shift behaviour. A dynamic balance between economy and society, intended to generate long-term relationships between user and object/ service and finally to be respectful and mindful of the environmental and social differences.[2]

Conceptual problems

Diminishing returns

The principle that all directions of progress run out, ending with diminishing returns, is evident in the typical 'S' curve of the technology life cycle and in the useful life of any system as discussed in industrial ecology and life cycle assessment. Diminishing returns are the result of reaching natural limits.

Common business management practice is to read diminishing returns in any direction of effort as an indication of diminishing opportunity, the potential for accelerating decline and a signal to seek new opportunities elsewhere.[citation needed] (see also: law of diminishing returns, marginal utility and Jevons paradox.)

Unsustainable Investment

A problem arises when the limits of a resource are hard to see, so increasing investment in response to diminishing returns may seem profitable as in the Tragedy of the Commons, but may lead to a collapse. This problem of increasing investment in diminishing resources has also been studied in relation to the causes of civilization collapse byJoseph Tainter among others.[3] This natural error in investment policy contributed to the collapse of both the Roman and Mayan, among others. Relieving over-stressed resources requires reducing pressure on them, not continually increasing it whether more efficiently or not[4]

Waste prevention

Negative Effects of Waste

About 80 million tonnes of waste in total are generated in the U.K. alone, for example, each year.[5] And with reference to only household waste, between 1991/92 and 2007/08, each person in England generated an average of 1.35 pounds of waste per day.[6]

Experience has now shown that there is no completely safe method of waste disposal. All forms of disposal have negative impacts on the environment, public health, and local economies. Landfills have contaminated drinking water. Garbage burned in incinerators has poisoned air, soil, and water. The majority of water treatment systems change the local ecology. Attempts to control or manage wastes after they are produced fail to eliminate environmental impacts.

The toxic components of household products pose serious health risks and aggravate the trash problem. In the U.S., about eight pounds in every ton of household garbage contains toxic materials, such as heavy metals like nickel, lead, cadmium, and mercuryfrom batteries, and organic compounds found in pesticides and consumer products, such as air freshener sprays, nail polish, cleaners, and other products.[7] When burned or buried, toxic materials also pose a serious threat to public health and the environment.

The only way to avoid environmental harm from waste is to prevent its generation. Pollution prevention means changing the way activities are conducted and eliminating the source of the problem. It does not mean doing without, but doing differently. For example, preventing waste pollution from litter caused by disposable beverage containers does not mean doing without beverages; it just means using refillable bottles.

Waste prevention strategies In planning for facilities, a comprehensive design strategy is needed for preventing generation of solid waste. A good garbage prevention strategy would require that everything brought into a facility be recycled for reuse or recycled back into the environment through biodegradation. This would mean a greater reliance on natural materials or products that are compatible with the environment.

Any resource-related development is going to have two basic sources of solid waste — materials purchased and used by the facility and those brought into the facility by visitors. The following waste prevention strategies apply to both, although different approaches will be needed for implementation:[8]

- Use products that minimize waste and are nontoxic
- Compost or anaerobically digest biodegradable wastes
- Reuse materials onsite or collect suitable materials for offsite recycling

Sustainable design principles

While the practical application varies among disciplines, some common principles are as follows:

- Low-impact materials: choose non-toxic, sustainably produced or recycled materials which require little energy to process
- Energy efficiency: use manufacturing processes and produce products which require less energy
- Emotionally Durable Design: reducing consumption and waste of resources by increasing the durability of relationships between people and products, through design
- Design for reuse and recycling: "Products, processes, and systems should be designed for performance in a commercial 'afterlife'."[9]
- Design impact measures for total carbon footprint and life-cycle assessment for any resource used are increasingly required and available.^ [10] Many are complex, but some give quick and accurate whole-earth estimates of impacts. One measure estimates any spending as consuming an average economic share of global energy use of 8,000 BTU (8,400 kJ) per dollar and producing CO2 at the average rate of 0.57 kg of CO2 per dollar (1995 dollars US) from DOE figures.
- Sustainable design standards and project design guides are also increasingly available and are vigorously being developed by a wide array of private organizations and individuals. There is also a large body of new methods emerging from the rapid development of what has become known as 'sustainability science' promoted by a wide variety of educational and governmental institutions.
- Biomimicry: "redesigning industrial systems on biological lines... enabling the constant reuse of materials in continuous closed cycles..."[12]
- Service substitution: shifting the mode of consumption from personal ownership of products to provision of services which provide similar functions, e.g., from a private automobile to a carsharing service. Such a system promotes minimal resource use per unit of consumption (e.g., per trip driven).[13]

- Renewability: materials should come from nearby (local or bioregional), sustainably managed renewable sources that can be composted when their usefulness has been exhausted.
- Robust eco-design: robust design principles are applied to the design of a pollution sources.[14]

Bill of Rights for the Planet

A model of the new design principles necessary for sustainability is exemplified by the "Bill of Rights for the Planet" or "Hannover Principles" - developed by William McDonough Architects for EXPO 2000 that was held in Hannover, Germany.

The Bill of Rights:

- Insist on the right of humanity and nature to co-exist in a healthy, supportive, diverse, and sustainable condition.
- Recognize Interdependence. The elements of human design interact with and depend on the natural world, with broad and diverse implications at every scale. Expand design considerations to recognizing even distant effects.
- Respect relationships between spirit and matter. Consider all aspects of human settlement including community, dwelling, industry, and trade in terms of existing and evolving connections between spiritual and material consciousness.
- Accept responsibility for the consequences of design decisions upon human well-being, the viability of natural systems, and their right to co-exist.
- Create safe objects of long-term value. Do not burden future generations with requirements for maintenance or vigilant administration of potential danger due to the careless creations of products, processes, or standards.
- Eliminate the concept of waste. Evaluate and optimize the full life-cycle of products and processes, to approach the state of natural systems in which there is no waste.
- Rely on natural energy flows. Human designs should, like the living world, derive their creative forces from perpetual solar income. Incorporating this energy efficiently and safely for responsible use.
- Understand the limitations of design. No human creation lasts forever and design does not solve all problems. Those who create and plan should practice humility in the face of nature. Treat nature as a model and mentor, not an inconvenience to be evaded or controlled.
- Seek constant improvement by the sharing of knowledge. Encourage

direct and open communication between colleagues, patrons, manufacturers and users to link long term sustainable considerations with ethical responsibility, and re-establish the integral relationship between natural processes and human activity.

These principles were adopted by the World Congress of the International Union of Architects (UIA) in June 1993 at the American Institute of Architects' (AIA) Expo 93 inChicago. Further, the AIA and UIA signed a "Declaration of Interdependence for a Sustainable Future." In summary, the declaration states that today's society is degrading its environment and that the AIA, UIA, and their members are committed to:

- Placing environmental and social sustainability at the core of practices and professional responsibilities
- Developing and continually improving practices, procedures, products, services, and standards for sustainable design
- Educating the building industry, clients, and the general public about the importance of sustainable design
- Working to change policies, regulations, and standards in government and business so that sustainable design will become the fully supported standard practice
- Bringing the existing built environment up to sustainable design standards.

In addition, the Interprofessional Council on Environmental Design (ICED), a coalition of architectural, landscape architectural, and engineering organizations, developed a vision statement in an attempt to foster a team approach to sustainable design. ICED states: The ethics, education and practices of our professions will be directed to shape a sustainable future.... To achieve this vision we will join... as a multidisciplinary partnership."

These activities are an indication that the concept of sustainable design is being supported on a global and interprofessional scale and that the ultimate goal is to become more environmentally responsive. The world needs facilities that are more energy efficient and that promote conservation and recycling of natural and economic resources.[15]

APPLICATIONS

Applications of this philosophy range from the microcosm — small objects for everyday use, through to the macrocosm — buildings, cities, and the Earth's physical surface. It is a philosophy that can be applied in the fields of architecture, landscape architecture, urban design, urban planning, engineering, graphic design, industrial design, interior design, fashion design and human-computer interaction.

Sustainable design is mostly a general reaction to global environmental crises, the rapid growth of economic activity and human population, depletion of natural resources, damage to ecosystems, and loss of biodiversity.[16] The limits of sustainable design are reducing.

Whole earth impacts are beginning to be considered because growth in goods and services is consistently outpacing gains in efficiency. As a result, the net effect of sustainable design to date has been to simply improve the efficiency of rapidly increasing impacts. The present approach, which focuses on the efficiency of delivering individual goods and services, does not solve this problem.

The basic dilemmas include: the increasing complexity of efficiency improvements; the difficulty of implementing new technologies in societies built around old ones; that physical impacts of delivering goods and services are not localized, but are distributed throughout the economies; and that the scale of resource use is growing and not stabilizing.

EXAMPLES

Emotionally durable design

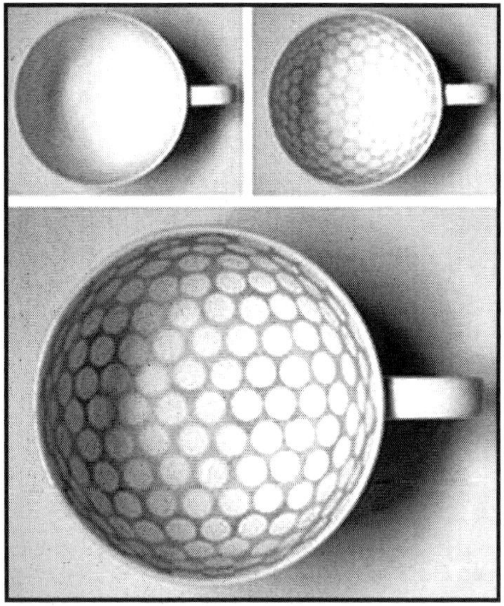

Fig. Stain Teacups: Bethan Laura Wood, 2009

According to Professor Jonathan Chapman of the University of Brighton, UK, emotionally durable design reduces the consumption and waste of natural resources by increasing the resilience of relationships established between consumers and products."[17] In his book, Emotionally Durable Design: Objects, Experiences & Empathy, Professor Chapman describes how "the process of consumption is, and has always been, motivated by complex emotional drivers, and is about far more than just the mindless purchasing of newer and shinier things; it is a journey towards the ideal or desired self, that through cyclical loops of desire and disappointment, becomes a seemingly endless process of serial destruction".[18]

According to Professor Chapman, 'emotional durability' can be achieved through consideration of the following five elements:

- Narrative: How users share a unique personal history with the product.
- Consciousness: How the product is perceived as autonomous and in possession of its own free will.
- Attachment: Can a user be made to feel a strong emotional connection to a product?
- Fiction: The product inspires interactions and connections beyond just the physical relationship.
- Surface: How the product ages and develops character through time and use.

As a strategic approach, "emotionally durable design provides a useful language to describe the contemporary relevance of designing responsible, well made, tactile products which the user can get to know and assign value to in the long-term."[19]According to Hazel Clark and David Brody of Parsons The New School for Design in New York, "emotionally durable design is a call for professionals and students alike to prioritise the relationships between design and its users, as a way of developing more sustainable attitudes to, and in, design things."[20]

Eco fashion and home accessories

Creative designers and artists are perhaps the most inventive when it comes to upcycling or creating new products from old waste. A growing number of designers upcyclewaste materials such as car window glass and recycled ceramics, textile offcuts from upholstery companies, and even decommissioned fire hose to make belts and bags. Whilst accessories may seem trivial when pitted against green scientific breakthroughs; the ability of fashion and retail to influence and inspire consumer behaviour should not be underestimated. Eco design may also use bi-products of industry, reducing the amount of waste being dumped in landfill, or may harness new sustainable materials or production techniques e.g. fabric made from recycled PET plastic bottles or bamboo textiles.

Sustainable architecture

Sustainable architecture is the design of sustainable buildings. Sustainable architecture attempts to reduce the collective environmental impacts during the production of building components, during the construction process, as well as during thelifecycle of the building (heating, electricity use, carpet cleaning etc.) This design practice emphasizes efficiency of heating and cooling systems; alternative energy sources such as solar hot water, appropriate building siting, reused or recycled building materials; on-site power generation - solar technology, ground source heat pumps, wind power; rainwater harvesting for gardening, washing and aquifer recharge; and on-site waste management such as green roofs that filter and control stormwater runoff. This requires close cooperation of the design team, the architects, the engineers, and the client at all project stages, from site selection, scheme formation, material selection and procurement, to project implementation.[21]

Sustainable architects design with sustainable living in mind.[22] Sustainable vs green design is the challenge that designs not only reflect healthy processes and uses but are powered by renewable energies and site specific resources.

A test for sustainable design is — can the design function for its intended use without fossil fuel — unplugged. This challenge suggests architects and planners design solutions that can function without pollution rather than just reducing pollution. As technology progresses in architecture and design theories and as examples are built and tested, architects will soon be able to create not only passive, null-emission buildings, but rather be able to integrate the entire power system into the building design. In 2004 the 59 home housing community, the Solar Settlement, and a 60,000 sq ft (5,600 m2) integrated retail, commercial and residential building, the Sun Ship, were completed by architect Rolf Disch inFreiburg, Germany. The Solar Settlement is the first housing community world wide in which every home, all 59, produce a positive energy balance.[23]

An essential element of Sustainable Building Design is indoor environmental quality including air quality, illumination, thermal conditions, and acoustics. The integrated design of the indoor environment is essential and must be part of the integrated design of the entire structure. ASHRAE Guideline 10-2011 addresses the interactions among indoor environmental factors and goes beyond traditional standards.[24]

Sustainable planning

Urban planners that are interested in achieving sustainable development orsustainable cities use various design principles and techniques when designing cities and their infrastructure. These include Smart Growth theory, Transit-oriented development, sustainable urban infrastructure and New Urbanism.

Smart Growth is an urban planning and transportation theory that concentrates growth in infill sites within the existing infrastructure of a city or town to avoid urban sprawl; and advocates compact, transit-oriented development, walkable, bicycle-friendly land use, including mixed-use development with a range of housing choices. Transit-oriented development attempts to maximise access to public transport and thereby reduce the need for private vehicles. Public transport is considered a form of Sustainable urban infrastructure, which is a design approach which promotes protected areas, energy-efficient buildings, wildlife corridors and distributed, rather than centralized, power generation and waste water treatment.

New Urbanism is more of a social and aesthetic urban design movement than a green one, but it does emphasize diversity of land use and population, as well as walkable communities which inherently reduce the need for automotive travel.

Both urban and rural planning can benefit from including sustainability as a central criterion when laying out roads, streets, buildings and other components of the built environment. Conventional planning practice often ignores or discounts the natural configuration of the land during the planning stages, potentially causing ecological damage such as the stagnation of streams, mudslides, soil erosion, flooding and pollution. Applying methods such as scientific modelling to planned building projects can draw attention to problems before construction begins, helping to minimise damage to the natural environment.

Cohousing is an approach to planning based on the idea of intentional communities. Such projects often prioritize common space over private space resulting in grouped structures that preserve more of the surrounding environment.

Watershed assessment of carrying capacity; estuary, riparian zone restoration and groundwater recharge for hydrologic cycle viability; and other opportunities and issues about Water and the environment show that the foundation of smart growth lies in the protection and preservation of water resources. The total amount of precipitation landing on the surface of a community becomes the supply for the inhabitants. This supply amount then dictates the carrying capacity - the potential population - as supported by the "water crop."

Sustainable landscape and garden design

Sustainable landscape architecture is a category of sustainable design and energy-efficient landscaping concerned with the planning and design of outdoor space. Plants and materials may be bought from local growers to reduce energy used in transportation. Design techniques include planting trees to shade buildings from the sun or protect them from wind, using local materials, and on-site composting and chipping not only to reduce green waste hauling but to increase organic matter and therefore carbon in thesoil.

Some designers and gardeners such as Beth Chatto also use drought-resistant plants in arid areas (xeriscaping) and elsewhere so that water is not taken from local landscapes and habitats for irrigation. Water from building roofs may be collected in rain gardens so that the groundwater is recharged, instead of rainfall becoming surface runoff and increasing the risk of flooding.

Areas of the garden and landscape can also be allowed to grow wild to encourage bio-diversity. Native animals may also be encouraged in many other ways: by plants which provide food such as nectar and pollen for insects, or roosting or nesting habitats such as trees, or habitats such as ponds for amphibians and aquatic insects. Pesticides, especially persistent pesticides, must be avoided to avoid killing wildlife. Soil fertility can be managed sustainably by the use of many layers of vegetation from trees to ground-cover plants and mulches to increase organic matter and thereforeearthworms and mycorrhiza; nitrogen-fixing plants instead of synthetic nitrogen fertilizers; and sustainably harvested seaweed extract to replace micronutrients.

Sustainable landscapes and gardens can be productive as well as ornamental, growing food, firewood and craft materials from beautiful places.

Sustainable landscape approaches and labels include organic farming and growing, permaculture, agroforestry, forest gardens, agroecology, vegan organic gardening and ecological gardening.

Sustainable graphic design

Sustainable graphic design considers the environmental impacts of graphic design products (such as packaging, printed materials, publications, etc.) throughout a life cycle that includes: raw material; transformation; manufacturing; transportation; use; and disposal. Techniques for sustainable graphic design include: reducing the amount of materials required for production; using paper and materials made with recycled, post-consumer waste; printing with low-VOC inks; and using production and distribution methods that require the least amount of transport.

Sustainable agriculture

Sustainable agriculture adheres to three main goals:

- Environmental Health,
- Economic Profitability,
- Social and Economic Equity.

A variety of philosophies, policies and practices have contributed to these goals. People in many different capacities, from farmers to consumers, have shared this vision and contributed to it. Despite the diversity of people and perspectives, the following themes commonly weave through definitions of sustainable agriculture.

There are strenuous discussions — among others by the agricultural sector and authorities — if existing pesticide protocols and methods of soil conservation adequately protect topsoil and wildlife. Doubt has risen if these are sustainable, and if agrarian reforms would permit an efficient agriculture with fewer pesticides, therefore reducing the damage to the ecosystem.

For more information on the subject of sustainable agriculture: "UC Davis: Sustainable Agriculture Research and Education Program".[25]

Domestic machinery and furniture

Automobiles, home appliances and furnitures can be designed for repair and disassembly (for recycling), and constructed from recyclable materials such as steel, aluminum and glass, and renewable materials, such as Zelfo, wood and plastics from natural feedstocks. Careful selection of materials and manufacturing processes can often create products comparable in price and performance to non-sustainable products. Even mild design efforts can greatly increase the sustainable content of manufactured items.

Improvements to heating, cooling, ventilation and water heating:

- Absorption refrigerator
- Annualized geothermal solar
- Earth cooling tubes
- Geothermal heat pump
- Heat recovery ventilation
- Hot water heat recycling
- Passive cooling
- Renewable heat
- Seasonal thermal energy storage (STES)
- Solar air conditioning
- Solar hot water.

Disposable products

Detergents, newspapers and other disposable items can be designed to decompose, in the presence of air, water and common soil organisms. The current challenge in this area is to design such items in attractive colors, at costs as low as competing items. Since most such items end up in landfills, protected from air and water, the utility of such disposable products is debated.

Energy sector

Sustainable technology in the energy sector is based on utilizing renewable sources of energy such as solar, wind, hydro, bioenergy, geothermal, and hydrogen. Wind energy is the world's fastest growing energy source; it has been in use for centuries in Europe and more recently in the United States and other nations. Wind energy is captured through the use of wind turbines that generate and transfer electricity for utilities, homeowners and remote villages. Solar power can be harnessed through photovoltaics, concentrating solar, or solar hot water and is also a rapidly growing energy source.[26]

The availability, potential, and feasibility of primary renewable energy resources must be analyzed early in the planning process as part of a comprehensive energy plan. The plan must justify energy demand and supply and assess the actual costs and benefits to the local, regional, and global environments.

Responsible energy use is fundamental to sustainable development and a sustainable future. Energy management must balance justifiable energy demand with appropriate energy supply. The process couples energy awareness, energy conservation, and energy efficiency with the use of primary renewable energy resources.[27]

Water sector

Sustainable water technologies have become an important industry segment with several companies now providing important and scalable solutions to supply water in a sustainable manner.

Beyond the use of certain technologies, Sustainable Design in Water Management also consists very importantly in correct implementation of concepts.

Among one of these principal concepts is the fact normally in developed countries 100% of water destined for consumption, that is not necessarily for drinking purposes, is of potable water quality.

This concept of differentiating qualities of water for different purposes has been called "fit-for-purpose".[28] This more rational use of water achieves several economies, that are not only related to water itself, but also the consumption of energy, as to achieve water of drinking quality can be extremely energy intensive for several reasons.

TERMINOLOGY

In some countries the term sustainable design is known as ecodesign, green design or environmental design. Victor Papanek, embraced social design and social quality and ecological quality, but did not explicitly combine these areas of design concern in one term. Sustainable design and design for sustainability are more common terms, including the triple bottom line (people, planet and profit).[citation needed]

SUSTAINABLE TECHNOLOGIES

Sustainable technologies use less energy, fewer limited resources, do not deplete natural resources, do not directly or indirectly pollute the environment, and can be reused or recycled at the end of their useful life.[29] There is significant overlap with appropriate technology, which emphasizes the suitability of technology to the context, in particular considering the needs of people in developing countries. However, the most appropriate technology may not be the most sustainable one; and a sustainable technology may have high cost or maintenance requirements that make it unsuitable as an "appropriate technology," as that term is commonly used.

DESIGN PRODUCTS FOR SUSTAINABILITY

The Bruntland Commission Report defined sustainable development as development that "meets the needs of the present without compromising the ability of future generations to meet their own needs." For a company to grow and secure its growth in the future, it needs to embed sustainability into all its products, services and processes.

Companies have long been looking for a way of quantifying sustainability and as a result Carbon Footprinting or Life-Cycle Analysis (LCA) have become commonplace approaches adopted to identify the impact of a company and its activities in terms of the environment. These are both appropriate as indicators of sustainability and involve calculating the embodied carbon within a product or activity and using this as a metric throughout the entire life-cycle of a product, service or process.

The basis of carbon footprinting and LCA stems from the idea of "Life-Cycle Thinking" which is, very simply, just looking at the life-cycle of a product, service or process from raw material extraction, through manufacture and distribution to ultimate disposal.

For everything that is manufactured, it makes sense to look at sustainability from the very beginning of the process, and thus the concept of eco-design has evolved over time.

Eco-design has been around for years, even Dieter Rams, chief designer at Braun in the 60's and 70's included environmental considerations in his 10 principles for good design. It can be described as a simple application of life-cycle thinking from a design perspective, and the benefits of doing so can include cost savings, legislative and regulatory compliance and customer satisfaction (or PR).

If a company wants to design a product with sustainability principles in mind, all it needs to do is to consider its eco-design and its life-cycle impacts and then minimise the biggest environmental impacts identified from this analysis. This is the first step to sustainable design.

The Design Process

During product or packaging design, the environmental impact should be considered at every stage in the life-cycle, from the raw material extraction through to the end of the product's life. Designers already do this when considering form or function; for example, a common design question is "how strong does packaging need to be to transport the product safely from the manufacturer to the consumer?". It is therefore only a small step for businesses to start to consider the life-cycle from a wider sustainability point of view.

For example, when manufacturing a mobile phone and looking at the consumer behaviour, we can see that it is often only used for twelve to eighteen months before it is replaced. Therefore, one of its biggest impacts would be disposal (which can be minimized by designing the phone for ease of recyclability), so a mobile phone company might want to examine the amount and mixture of materials from which it is made to help minimize any impacts associated with its dismantling and disposal. During the 'eco-design' process the company would need to consider its manufacturing using as little (and as few) materials as possible. If the phone is compared to the mobile phone charger, the biggest environmental impact of this is almost definitely the amount of energy expended during its usage. It would therefore be sensible to 'eco-design' the phone charger by trying to optimise the energy efficiency during usage.

The Design Council recently estimated that 80 % of the cost of a product is set at the design stage, and therefore reducing the environmental impact of any product during the concept design is actually the most beneficial stage at which to make cost savings.

Eco-Design – Tools and Techniques

There are a number of tools and techniques that can be used to design products more sustainably, and the right technique will depend on each company's aims and objectives. For example, if a company is looking to reduce its carbon footprint, then it would make sense to look at "Design for embedded carbon" and review the material selection, or look at "Design for transport efficiency", as the distribution of the product may well cause the biggest production of carbon.

However, if a company has set targets for moving to 100% recyclable packaging, then it would need to look at "Design for recyclability" and move towards using mono materials that can easily be separated at point of disposal and recycled in most local authorities' collection streams. Companies need to be careful, however, when transporting packaging or products abroad that the materials can be readily recycled at their destination. A few of the techniques commonly used for minimizing environmental impact are outlined below.

Packaging eco-design techniques

Design for embedded carbon

- Look at the material used in the product or its packaging; for example, using Aluminum that is made from 60% recycled content can reduce the product's embedded carbon by up to 90%

Design for recyclability

- Consider the recyclability of the materials from which the product or packaging is made

- Minimize the different types of materials used and, if possible, move to a single material product
- Look at how the materials are fixed together; for example, moving from screws to snap clips reduces the amount of time it takes to dismantle the product and they could also be made from the same material

Design for recycled content

- Most modern materials can include high levels of recycled content, for example cardboard boxes, metals and most plastics. An obvious and commonly-used example is the Innocent Drinks bottle, one of the first to be made from 100% recycled PET
- By asking suppliers for more recycled content in the materials purchased, costs can often be cut and money can be saved

Design for bio-degradability or compostability

- Does the consumer have the ability to compost? If so, moving to biodegradable packaging (which is suitable for home composting) can minimize the impact of the packaging at the end of its life
- However, care must be taken and the company needs to ensure that the packaging really will be composted. The EU Landfill Directive sets demanding targets to reduce the amount of biodegradable municipal waste going to landfill, one of the reasons being this type of material can increase methane and CO2 production by up to 20 times!

Design for transport efficiency

- Can the packaging be designed so that more products fit onto one pallet?
- Can the packaging be designed to interlock or stack in a different way to allow more products to stack together?
- Can shelf-ready packaging be introduced, thus eliminating the need for secondary and transit packaging and therefore fitting more products together in one pack?

Product eco-design techniques

Design for concentration

- If a product contains water, for example cleaning products, paints, coatings or drinks, can it be concentrated so the consumer can mix it with water at it's destination? This means smaller (and cheaper) packaging, lower transport and storage costs and sometimes a longer lifespan of the product

Design for longevity

- Historically, some companies have been accused of planned obsolescence, which is deliberately planning or designing a product with a limited useful life, so that it will become obsolete or nonfunctional after a certain period to ensure consumers re-purchase products
- Most designers are, however, now moving away from inbuilt obsolescence and looking at whether the product can be designed to last longer, for example a kitchen knife with 2 blades, so that, once the user cannot re-sharpen the first blade satisfactorily, the blade can be swapped and the blunt one sent back to the manufacturer to be professionally sharpened. Another example is that of a washing re-programmable machine, so that when a new washing powder is released that allows consumers to wash at a lower temperature, a new programme can be uploaded that sets the temperature to the new level

Design for energy efficiency

- Products that use energy are starting to be covered by new regulations (under the European Energy Using Products Directive) which set out eco-design requirements, mostly to do with energy efficiency in use. Therefore, manufacturers are starting to have to document and reduce the energy used in standby, on and powered-down modes

All of the above can (and should) be considered during the design stage of any product or packaging. A good way to do this is to undertake a workshop, inviting representatives from all the different sections of the business, from marketers, production managers and environmental managers to the senior management to attend and contribute. Brainstorming with these different staff together, looking at product lines as specific examples and building short, medium and long term plans for improvements, quite often identifies projects where low cost/ no cost changes can save vast amounts of money. It is worth remembering that, although external consultants can often add value by providing additional advice and expertise and by helping to facilitate the workshop discussion, no-one knows a company better than its own staff!

Green Consumption

The green consumer market grew by 15% in 2008, whereas the overall figure for the consumer market growth was nearer 1.4%, with estimates on sustainable food up by 14%, sustainable textiles up 71%, green stationery up 49% and even eco-friendly funerals up by 18%, now is the ideal time for companies to grow by producing and marketing more sustainable products.

Index

O

P

S

U